TROPICAL AND PARASITIC INFECTIONS IN THE INTENSIVE CARE UNIT

PERSPECTIVES ON CRITICAL CARE
INFECTIOUS DISEASES
Jordi Rello, M.D., Series Editor

TROPICAL AND PARASITIC INFECTIONS IN THE INTENSIVE CARE UNIT

edited by

Charles Feldman, MB BCh., PhD, FRCP, FCP (SA).
Division of Pulmonology,
Department of Medicine,
Johannesburg Hospital, and
University of the Witwatersrand,
Johannesburg, South Africa.

and

George A. Sarosi, M.D., M.A.C.P.
Indiana University School of Medicine,
and Medical Service,
Veterans Administration Medical Center,
Indianapolis, USA

 Springer

Charles Feldman, MB BCh., PhD, FRCP, FCP (SA)
Division of Pulmonology
Department of Medicine
Johannesburg Hospital and
University of the Witwatersrand
Johannesburg, South Africa

George A. Sarosi, M.D., M.A.C.P.
Indiana University School of Medicine
and Medical Service
Veterans Administration Medical Ctr.
Indianapolis, Indiana, U.S.A.

Library of Congress Cataloging-in-Publication Data

Tropical and parasitic infections in the intensive care unit / edited by Charles Feldman and
    George A. Sarosi.
        p. cm. – (Perspectives on critical care infectious diseases ; 9)
    Includes bibliographical references and index.
    ISBN 0-387-23379-2   Printed on acid-free paper.   e-ISBN 0-387-23380-6
        1. Tropical Medicine. 2. Critical care medicine. 3. Communicable diseases. 4. Intensive
care units. I. Feldman, Charles, 1952- II. Sarosi, George A. III. Series.

RC961.T734 2005
616.9'883-dc22   2004058994

Printed in the United States of America.

9 8 7 6 5 4 3 2 1   SPIN 11051466

springeronline.com

CONTENTS

v

CONTRIBUTORS

Christine L. N. Banage, MB ChB, FCP (SA).
Intensive Care Department
Chris Hani Baragwanath Hospital,
Johannesburg, SOUTH AFRICA

Lucille Blumberg, MB BCh, MMED (Med Micro), DTM&H, DOH, DCH.
National Institute for Communicable Diseases,
Johannesburg, SOUTH AFRICA

Loren C. Denlinger, M.D., PhD.
Division of Pulmonary and Critical Care,
Department of Medicine,
University of Wisconsin-Madison,
and the University of Wisconsin Hospital and Clinics,
Madison, Wisconsin, USA

Scott E. Evans, M.D.
Division of Pulmonary, Critical Care Medicine,
and Internal Medicine,
Mayo Clinic and Foundation,
Rochester, Minnesota, USA

Charles Feldman, MB BCh, PhD, FRCP, FCP (SA).
Division of Pulmonology,
Department of Medicine
University of the Witwatersrand
Johannesburg, SOUTH AFRICA

John Frean, MB BCh, MMED (Path).
National Institute for Communicable Diseases,
National Health Laboratory Services,
Johannesburg, SOUTH AFRICA

Jeffery Glassroth, M.D.
Division of Pulmonary and Critical Care,
Department of Medicine,
University of Wisconsin-Madison,
and the University of Wisconsin Hospital and Clinics,
Madison, Wisconsin, USA

David E. Greenberg, M.D.
NIAID, Laboratory of Infectious Diseases,
Bethesda, Maryland, 20892, USA

Stephen B. Greenberg, M.D.
Department of Medicine,
and Department of Molecular Virology and Microbiology,
Baylor College of Medicine,
Houston, Texas, 77030, USA

Chadi A. Hage, M.D.
Indiana University-School of Medicine,
Indianapolis, IN, USA

Alan S. Karstaedt, MB BCh, MMED (Int Med), DTM&H.
Division of Infectious Diseases
Department of Medicine,
Chris Hani Baragwanath Hospital
University of the Witwatersrand,
Johannesburg, SOUTH AFRICA

Kenneth S. Knox, M.D.
Indiana University-School of Medicine,
and Richard L. Roudebusch VA Medical Center,
Indianapolis, IN, USA

Linda S. Lewis, D.V.M., M.P.V.M.
Health Studies Consulting
Medford, OR 97501, USA

Andrew H. Limper, M.D.
Division of Pulmonary, Critical Care Medicine
and Internal Medicine,
Mayo Clinic and Foundation,
Rochester, Minnesota, USA

L. Rudo Mathivha, MB ChB, FCP (Critical Care).
Intensive Care Department
Chris Hani Baragwanath Hospital,
Johannesburg, SOUTH AFRICA

Mervyn Mer MB BCh, MMED (Int Med), FCP (SA).
Division of Pulmonology,
Department of Medicine
University of theWitwatersrand
Johannesburg, SOUTH AFRICA

Monica E. Parise, M.D.
Malaria Epidemiology Branch
Division of Parasitic Diseases
National Center for Infectious Diseases
Center for Disease Control and Prevention
Atlanta, GA 30341-3717, USA

Guy A. Richards MB BCh, PhD, FRCP, FCP (SA).
Intensive Care Unit,
Johannesburg Hospital
Department of Medicine,
University of the Witwatersrand,
Johannesburg, SOUTH AFRICA

Ian M. Sanne MMB BCh, FCP (SA), DTM&H.
Clinical HIV Research Unit
University of the Witwatersrand,
Johannesburg, SOUTH AFRICA

George A. Sarosi, M.D., M.A.C.P.
Indiana University-School of Medicine,
and Roudebusch VA Medical Center,
Indianapolis, IN, USA

Gunter Schleicher MB BCh, MMED (Int Med), FCP (SA).
Division of Pulmonology,
Department of Medicine
University of theWitwatersrand
Johannesburg, SOUTH AFRICA

W. D. Francois Venter MB BCh, FCP (SA), DTM&H.
Reproductive Health Research Unit,
University of the Witwatersrand,
Johannesburg, SOUTH AFRICA

Hayden T. White, MB BCH, MMED (Int Med), FCP (SA).
Division of Pulmonology,
Department of Medicine
University of the Witwatersrand
Johannesburg, SOUTH AFRICA

PREFACE

The term *tropical infection* is most commonly used to describe an infection occurring in an individual living in the geographical area between the Tropic of Cancer and the Tropic of Capricorn. Both developing and developed countries fall within these equatorial parallels. It is said that the common feature that allows the specific pathogens to flourish in these areas is the hot and humid climate. While a myriad of different micro-organisms may cause tropical infections, many of these infections are caused by a relatively small number of common bacteria, mycobacteria, viruses, fungi and parasites. Several of these infections may also cause critical illness. Human immunodeficiency virus (HIV) infection is also found commonly in many of these tropical countries and so there is considerable overlap between HIV infection and tropical infections.

In the past many of these infections may have been confined to these tropical areas of the world. However, with the increase in air travel and tourism and the changing patterns of immigration, an increasing number of individuals are coming into contact with these infectious agents and transmission across the world has been enhanced. Such infections are therefore being seen more often and treated in apparently unusual locations. For these reasons this reference volume entitled "Tropical and Parasitic Infections in the Intensive Care Unit" provides an important overview of those infections that may cause critical illness. A unique aspect of this series of volumes is the individual reviews given by authors from different parts of the world, imparting their own perspectives to each of the chapters.

Charles Feldman
George Sarosi
Book Editors

1

Severe Malaria

Manifestations, diagnosis, chemotherapy, and management of severe malaria in adults

Lucille Blumberg
National Institute for Communicable Diseases, South Africa

INTRODUCTION

The burden of malaria is increasing, especially in sub-Saharan Africa, because of drug and insecticide resistance and social and environmental changes (1). Each year an estimated three to four hundred million people will contract malaria globally, resulting in five hundred thousand to two million deaths. Ninety percent of the world's malaria, and at least 90% of malaria-related mortality, occurs in Sub-Saharan Africa, primarily in young children (2). Malaria occurs in every country in sub-Saharan Africa, with the exception of Lesotho, but transmission rates vary within regions and within countries. In parts of Africa where endemicity of malaria is high and transmission stable, such as Tanzania, Malawi, and Mozambique, severe malaria is mainly a disease of children under 5 years of age and of pregnant women. It is less common in older children and adults because of the partial immunity acquired as a result of repeated infections. In areas of low endemicity severe malaria occurs in both adults and children. Non-immune travellers to malaria areas are always at risk for severe disease (3,4).

The majority of malaria cases in Africa are due to *Plasmodium falciparum*, the major species associated with mortality and morbidity. The development of parasite resistance to chemotherapeutic agents such as chloroquine has resulted in a significant increase in malaria morbidity and mortality. The demise of chloroquine, an affordable option in resource-poor countries, has major implications for malaria management (5). In Africa resources for management of severe malaria are limited and at least 20-30% of patients with complications of disease will die. In a confidential inquiry into malaria deaths in an area of South Africa with limited tertiary care facilities, major contributing factors were delays in diagnosis and initiation of adequate therapy, failure to administer the correct antimalarial at the correct dosage and frequency, inadequate monitoring of severity indicators in complicated cases, and the sub-optimal management of complications (6).

PATHOGENESIS OF SEVERE MALARIA

Key features of malaria are the adherence of infected red blood cells to the endothelium of small blood vessels compromising blood flow through tissues, and the production of pro-inflammatory cytokines (7). Factors that determine whether a patient develops mild or severe disease are complex and multifactorial and are related to both the parasite and the host. Parasites causing severe malaria have a greater multiplication potential than those causing uncomplicated infections (8). The effect of inoculum dose on severity is unclear and difficult to investigate. Cyto-adherence of parasitised red cells may be influenced by the virulence of different strains of parasite (9).

The development of immunity to the clinical effects of malaria requires several years of continuous exposure. Lack of this protective immunity would be expected to be the major factor determining the severity of a clinical attack of malaria. Differences in HLA antigens may play a role in host predisposition to severe disease. Certain red blood cell abnormalities, including sickle-cell trait, protect against malaria disease. Prevalence rates of these abnormalities are high in some parts of Africa and may provide some protection against severe malaria (9). Plasma interleukin (IL-6, IL-10) and tumour necrosis factor-α and the IL-6 : IL-10 ratio is significantly higher in patients who die than in survivors (10).

CLINICAL MANIFESTATIONS AND DIAGNOSIS

Symptoms and signs of malaria may present as early as seven days, but more commonly an average of 10-21 days after being bitten by an infected mosquito. Fever is prominent, but may be absent in some cases. Some of the following symptoms may also appear: rigors, headache, myalgia, diarrhoea, vomiting and cough. Physical signs may include fever, anaemia, jaundice, hepatosplenomegaly and a variety of cerebral signs. Malaria should be suspected in any person presenting with any of the above symptoms or signs with a history of travel to, or residence in a malaria transmission area. Presentation is very variable and may mimic other diseases, including influenza, hepatitis, meningitis, septicaemia, typhoid, tickbite fever, viral haemorrhagic fever, trypanosomiasis, HIV seroconversion illness, and relapsing fever (4).

P. falciparum infections may progress rapidly to a lethal, multi-system disease. The diagnosis of malaria is urgent, and complications can develop rapidly within 48 hours of the onset of disease in any non-immune person but especially in young children and pregnant women (4). The clinical manifestations of severe malaria depend on the age of the patient. In children, hypoglycaemia, convulsions, and severe anaemia are relatively common; acute renal failure, jaundice, and ARDS are more common in adults. Cerebral malaria, shock and acidosis may occur at any age (11). A number of clinical and laboratory criteria are used to define severe malaria, as shown in Table 1 (4,11).

Table 1: Indicators of severe malaria

Clinical Features
- Impaired consciousness, convulsions
- Respiratory distress: acidosis, ARDS, pulmonary oedema
- Jaundice
- Bleeding
- Shock

Biochemical Features
- Renal impairment – serum creatinine >265µmol/L, rapidly rising creatinine or urine output <400 ml/day (adult)
- Acidosis (plasma bicarbonate <15 mmol/L) (serum lactate >5 mmol/L)
- Hepatic impairment (transaminases >3 times normal)
- Hypoglycaemia (blood glucose <2.2 mmol/L)
- Hypoxia (PO_2 - <8 Kpa in room air)

Haematological Features
- Parasitaemia ≥5% or ≥3+
- Haemoglobin <6 g/dL or haematocrit <20%
- ≥5% neutrophils contain malaria pigment
- Presence of schizonts of *P.falciparum* in peripheral blood smear
- Evidence of DIC

Laboratory diagnosis

Patient blood should be examined immediately to confirm or exclude the diagnosis of malaria. In the majority of cases of severe malaria, examination of correctly stained blood smears will reveal malaria parasites, however, a negative smear does not exclude the diagnosis, and repeat smears are indicated. Some patients with severe malaria may have a negative smear due to sequestration of parasitised red blood cells, and a decision to treat with antimalarial chemotherapy should be considered if the index of suspicion is very high. In these cases it is imperative to continue to look for alternative diagnoses, especially trypanosomiasis, septicaemia and viral haemorrhagic fever.

High levels of parasitaemia ($\geq 5\%$) are generally predictive of severe malaria in nonimmune patients. Importantly, the converse may not be true, with severe disease also occurring with low parasitaemias in the peripheral blood (11,12). Quantification is often inaccurate, peripheral parasitaemia may not reflect the total parasite load and sequestration in the organs, and levels of parasitaemia may vary cyclically. Prognosis worsens considerably if *P. falciparum* schizonts are present in a blood smear, and if more than 5% of peripheral polymorphonuclear leucocytes contain visible malaria pigment (13).

Commercial kits are available that rapidly detect parasite antigen or enzymes. The tests for *P. falciparum* are highly sensitive, but depend on correct usage, interpretation of results, and the quality of the particular test used. These tests can only be used for diagnosis of acute malaria infections, and not for follow-up, as the test may remain positive for several weeks, even after successful treatment (14).

In a febrile patient where there is no obvious cause of fever, and a recent history of visiting or living in a malaria area is not forthcoming, malaria should still be excluded, as infected mosquitoes have been documented to travel long distances in road, rail and air transport. Mortality is high in this group of patients, because of missed diagnosis, but a finding of thrombocytopenia should always stimulate a search for possible malaria parasites (15).

TREATMENT OF SEVERE MALARIA

Patients should be treated urgently with the most effective treatment regimen available, in a facility with the highest level of care. The choice of chemotherapy for malaria is dependent on the severity of disease, the known or suspected resistance pattern of the parasite in the area where the malaria infection was acquired, the species of parasite, and patient profile (age, pregnancy, comorbidity, allergies, and medications, including any antimalarials recently administered).

Quinine, the drug of choice for the treatment of severe malaria in Africa, is rapidly effective (4,11). In most parts of Africa quinine resistance has not developed. In some parts of West Africa however, foci of low-level resistance have been documented (15). An initial loading dose of quinine to rapidly reach a therapeutic level is critical in the management of severe malaria and has a major impact on favourable outcome. The loading dose should be omitted if the patient has definitely received mefloquine, quinine, quinidine or halofantrine in the previous 24 hours, mefloquine in the previous seven days, or 40mg/kg of quinine in the previous two days. If there is doubt, the loading dose of quinine should be given (4,11,16). The loading dose is given as quinine di-hydrochloride salt, 20mg/kg body weight diluted in 5-10 ml/kg body weight of dextrose water, by slow intravenous infusion over two to four hours. Quinine must never be administered by bolus intravenous injection, as this is associated with cardiotoxicity. The loading dose is given strictly according to body weight. The disposition of quinine in very obese patients is not known. It has been suggested that there is a ceiling dose above which quinine should not be given, but there is no evidence to support this (17).

Six to eight hours after starting the loading dose, a maintenance dose of quinine di-hydrochloride salt, 10mg/kg diluted in 5-10 ml/kg body weight of a dextrose-containing solution should be commenced and infused over 4-6 hours. Intravenous quinine should be administered every eight hours until the patient can take oral medication (usually by 48 hours). For obese patients, the maintenance dose should be calculated according to ideal body weight (17).

Males: IBW (kg) = 0.9 x height in cm − 88
Females: IBW (kg) = 0.9 x height in cm − 92.

The dosage of oral quinine is 10mg/kg/dose or 600mg/dose given three times a day. The total duration of quinine therapy is 7-10 days. Additional drugs, tetracycline (usually as doxycycline 100mg twice a day x 7 days), or clindamycin (10mg/kg twice a day x 7 days) are recommended to improve cure rates (4,11,18). These, however, do not add initial therapeutic benefit, may contribute to drug side effects, and should be introduced only once the patient is improving. Quinine can be administered by deep intramuscular injection if intravenous infusion is not possible (4).

Quinine has a narrow therapeutic window, although serious side effects are rare. The pharmacokinetic properties of quinine are altered considerably in malaria with a contraction in the volume of distribution and a reduction in clearance that is proportional to the severity of disease (19). There is significant binding of quinine to acute phase reactants, notably α1-acid glycoprotein, with reduction in the levels of free quinine. Quinine toxicity is, therefore, relatively uncommon (20). The most frequent side effect of quinine therapy is hypoglycaemia, especially in children and pregnant women (19,21). Although quinine may prolong the QTc-interval, hypotension, heart block, and ventricular arrhythmias are rare (4,19,22). Convulsions and visual disturbances have been reported as idiosyncratic responses or with overdosage (4,19). Doses should be reduced by 30-50% after the third day of treatment to avoid accumulation of the drug in patients who remain seriously ill, especially those with evidence of renal failure (4). The measurement of levels of free (not total) quinine may be helpful in patients with severe malaria and renal failure, but accessibility to this test is very limited. The precise level has not been defined but probably lies between 0.8-2mg/L (11).

Quinidine is more active than quinine, but is also more cardiotoxic and more expensive, is not readily available, and consequently is not used for treating severe malaria in Africa (23).

Artemisinins
In the early 1970's Chinese scientists identified artemisinin, a sesquiterpene lactone peroxide, as the principal active component of the traditional Chinese malaria remedy, qinghaosu. Artemisinin and two derivatives, artesunate and artemether are effective against multi-drug resistant *P falciparum* and clear sensitive parasites from the blood more rapidly than other antimalarial agents due to their broad stage specificity of anti-malarial action. Despite administration to over 3 million people, resistance has not emerged, and only rarely has treatment failure been reported. The drugs are well tolerated and despite neurotoxicity in animal studies, serious adverse reactions have included only a few case reports of anaphylaxis. The chemical structure and mode of action of these drugs distinguish them from other currently available antimalarial agents, and render them less vulnerable to cross-resistance. However, when used alone, unacceptably high recrudescence rates are seen (4,11,24,25).

Combination therapy, which includes an artemisinin, is the recommended malaria treatment policy to delay the emergence of drug resistance to sequential monotherapy, as well as to improve cure rates. Drugs used in combination with the artemisinins include mefloquine, sulfadoxine pyrimethamine, amodiaquine and lumefantrine, and the choice depends on parasite resistance in the geographical area (26).

There are parenteral preparations of the artemisinins, either intramuscular (artemether, arteether, artesunate) or intravenous (artesunate). Artemether and arteether are oil-based preparations and absorption from the intramuscular site may be compromised in severe malaria, leading to treatment failures (27). Artesunate is water-based, can be given intravenously, or intramuscularly from where it is well absorbed. Although theoretically preferable, there are no large comparative trials to indicate whether artesunate is superior to artemether or quinine (28). The use of parenteral artemisinins is limited by availability and manufacturing practices, which may not adhere to international standards.

A meta-analysis of randomized clinical trials comparing the efficacy of artemether with quinine in the management of severe malaria demonstrated equality, but indicated a trend toward greater effectiveness of artemether in regions where there is recognised quinine resistance. Artemether was superior to quinine in terms of overall serious adverse events (29,30). In patients with hyperparasitaemia there may be an advantage of the artemisins over quinine. In South-East Asia, where multi-drug-resistant malaria is a major problem and quinine resistance has emerged, the artemisinin drugs are used as first-line therapy for severe malaria (30).

Other drugs

Widespread, high-level chloroquine resistance precludes the use of chloroquine in the treatment of both uncomplicated and severe malaria in most parts of the world, including Africa. Sulfadoxine pyrimethamine, mefloquine and halofantrine are not indicated in the management of severe malaria (4).

COMPLICATIONS OF MALARIA AND THEIR MANAGEMENT

Anaemia

Anaemia may result from haemolysis or dyserythropoeisis (31). Severe anaemia is defined as a haemoglobin of less than 6g/dL, or haematocrit <20%. Severe anaemia is the most important manifestation of severe malaria in areas of high stable transmission and occurs predominantly in children. Pregnant women may also present with profound degrees of anaemia. Anaemia may manifest as shock, cardiac failure, hypoxia, or confusion and the rate at which anaemia develops is an important determinant of compensatory mechanisms. Blood transfusion using packed cells should be considered in patients in whom the haemoglobin is 6g/dL or less, especially those with cardiovascular decompensation. Fluid overload must be avoided. Transfused blood has a reduced lifespan in malaria patients (4).

Cerebral malaria

In many parts of the world cerebral malaria is the most common clinical presentation and cause of death in adults with severe malaria. The term cerebral malaria in many published studies is restricted to the syndrome in which altered consciousness associated with a malaria infection could not be attributed to convulsions, sedative drugs or hypoglycaemia alone or to a non-malarial cause. Cerebral malaria may be part of multi-system pathology, in which case the outlook is much poorer than if disease was localised only to the central nervous system.

Clinically, the commonest neurological picture is of a symmetrical upper motor neuron lesion, mild neck stiffness is not uncommon, and muscle tone and tendon reflexes are variable. Cerebral malaria can resemble bacterial or viral meningitis and a lumbar puncture should be considered in patients where the diagnosis is not clear. Hypoglycaemia, metabolic disturbances, severe anaemia and hypoxia as a result of malaria can all present with signs of central nervous system dysfunction. Generalised or focal convulsions may occur as a result of cerebral malaria, or in association with hypoglycaemia (4).

Imaging of the brain commonly shows evidence of mild cerebral swelling. Oedema is very unusual, and may be an agonal phenomenon (32,33). Studies to date with dexamethasone or mannitol have not shown benefit and have been associated with prolongation of coma and gastro-instestinal haemorrhage (34). Anticonvulsants should only be used once convulsions occur, and should not be used prophylactically (35). The use of iron-chelating agents has not been shown to impact on mortality (36). In adult patients who recover, neurological sequelae are uncommon.

Renal failure

Renal failure is an early complication of severe malaria in adults. Hypovolaemia, sequestration of parasitised red cell in the renal vasculature, intravascular haemolysis and haemoglobinuria are implicated and may lead to acute tubular necrosis. Renal failure is generally oligaemic and hypercatabolic. A serum creatinine of greater than 256 μmol/L or a rapidly rising creatinine and/or a urine output of < 400 ml/day in an adult should be regarded as renal failure. A central venous catheter (CVP) should be inserted and dehydration should be corrected. The CVP should not be above 5cm of water. The indications for dialysis are the same as for patients with other diseases, but since renal failure in malaria occurs against a background of a hypercatabolic state and non-renal causes of acidosis frequently co-exist, early dialysis is recommended (37). Venovenous haemofiltration is the recommended mode of dialysis and is significantly more efficient than peritoneal dialysis (38).

Quinine is not removed by dialysis and in patients with severe malaria and renal failure, the dosage of quinine should be reduced by half to one-third after 2 days of full dosage administration. If the patient survives the acute phase of the disease and has no pre-existing underlying disease, recovery of renal function generally occurs within three weeks (4).

Respiratory failure

This is a grave complication of severe falciparum malaria in adults, and may present several days after commencing malaria chemotherapy. The cause of this often lethal complication is unknown in falciparum malaria. Some cases show evidence of pulmonary oedema while others resemble acute respiratory distress syndrome. Pregnant women are particularly at risk. Iatrogenic overadministration of fluids may contribute to the development of ARDS or pulmonary oedema and should be avoided. Some patients may require ventilatory support (4,11,39,40).

Hepatic dysfunction

Although a raised indirect bilirubin due to haemolysis is a frequent finding in malaria, the clinical presence of jaundice or the finding of raised hepatic transaminases (≥3 x normal) should alert the clinician to the probability of severe malaria. The presence of jaundice combined with renal failure and acidosis may indicate a grave prognosis (4).

Disseminated intravascular coagulation (DIC)

DIC is rare in patients with severe malaria although laboratory evidence of haemostatic abnormalities may be present without bleeding. Moderate degrees of thrombocytopenia are noted in the majority of cases of uncomplicated malaria unassociated with other coagulation abnormalities and bleeding is uncommon. Possible mechanisms of thrombocytopenia include sequestration in the spleen, decreased production, or reduced survival from intravascular lysis. Platelet transfusion should be considered if the platelet count is less than 20,000/mm³, or if there is evidence of bleeding. Platelet counts should return to normal within a few days with effective malaria treatment. Continuing thrombocytopenia may indicate failed antimalarial therapy, sepsis, or a drug reaction to quinine (4)

Secondary infection

Secondary bacterial infections may complicate malaria: aspiration pneumonia, urinary tract infections or nosocomial septicaemia. In a significant number of patients, especially children, septicaemia may complicate severe malaria very early. Salmonella species and staphylococci are common causes of septicaemia. The syndrome is associated with high mortality. Since the features of bacterial sepsis and malaria overlap, empiric treatment using a broad-spectrum antibiotic for Gram-positive and Gram-negative organisms is recommended (4).

Acidosis

Metabolic acidosis is a consistent feature of severe malaria. Lactic acidosis is a major cause of death from severe falciparum malaria. The pathophysiology of acidosis is multifactorial and results from tissue hypoxia and anaerobic glycolysis, liver dysfunction and impaired renal handling of bicarbonate. The presence of acidosis is an important predictor of poor outcome (41). The management of acidosis includes correction of fluid balance, improvement in haemodynamic status, and haemodialysis (4). The use of dichloracetate has been shown to be beneficial in animal models.

Haemoglobinuria and Blackwater fever

The pathogenesis of this rare condition is unknown, and is seen in patients with G-6-PD deficiency who receive oxidant drugs. It may also occur in patients without apparent G-6-PD deficiency but who have severe malaria and are treated with quinine or artemisinin derivatives. Intravascular haemolysis results in anaemia, and the passage of haemoglobinuria. A small minority will develop renal failure, the cause of which is unknown. In patients with malarial haemoglobinuria, quinine chemotherapy should be continued. Supportive therapy includes blood transfusions for severe anaemia, maintaining adequate hydration, and renal dialysis where indicated (4,42).

Hypoglycaemia

Hypoglycaemia may result from impaired glycolysis or gluconeogenesis, or as a result of quinine-induced hyperinsulinaemia. It is a particular problem in pregnant women and patients on intravenous quinine. Blood glucose should be monitored, as the signs may be very subtle. Hypoglycaemia must be excluded in all patients with an altered mental state and in those who present with convulsions (4).

Shock

Shock may occur as a result of hypovolaemia, massive blood loss from splenic rupture or gastrointestinal haemorrhage, bacterial septicaemia, hypoxia and severe metabolic acidosis. Myocardial function is remarkably good in severe falciparum malaria and most patients have an elevated cardiac index (43). Hypovolaemia should be corrected with an appropriate intravenous infusion, usually 0.9% saline initially, followed by a plasma expander. The central venous pressure should not be allowed to exceed 5cm. If hypotension persists, inotropes should be administered (4).

Pregnancy

The placenta acts as a haven for parasites due to upregulation of adhesion receptors. The course of malaria in pregnancy is rapidly progressive and common complications are anaemia, hypoglycaemia and ARDS. The risk of severe disease extends into the immediate postpartum period. Malaria may cause abortion, premature delivery and low birth- weight. The management remains the same as in non-pregnant patients, with emphasis on preventing and managing the complications mentioned. In particular, fluid restriction is important to prevent ARDS. Quinine is the drug of choice but may be associated with intractable hypoglycaemia. The use of the artemisinin drugs is currently not indicated due to a lack of safety data, unless there is evidence of quinine resistance. There is no indication to terminate pregnancy. In areas of high malaria transmission, anaemia is the most common manifestation of severe disease and placental parasitaemia is associated with low birth-weight infants (4).

Non-falciparum malaria

The non-falciparum malarias are not generally associated with severe disease due to a lack of sequestration of parasitised red cells. Rarely *Plasmodium vivax* has been associated with the development of ARDS and cerebral malaria (44,45). Mixed infections with falciparum malaria occur occasionally and should be managed as for falciparum malaria.

HIV and severe malaria

Malaria and human immunodeficiency virus (HIV) infections are common, widespread and overlapping problems in Africa. Any interaction between these two pandemics would be of great importance. This interaction could be in either direction, with malaria causing more rapid progression of HIV, and HIV-associated immunosuppression leading to an impaired immune response to malaria. Greater parasite densities and rates of clinical malaria have been demonstrated in HIV-positive patients from Uganda, an area of high malaria endemicity, where the majority of people would be expected to have developed some malaria immunity (46,47). In a cohort study of non-immune patients with malaria in South Africa, HIV-positive patients had an increased rate of severe malaria compared to HIV-negative patients, and the rate increased as CD4+ cell count decreased. HIV-positive patients had significantly increased rates of renal failure, severe anaemia and DIC (48).

Exchange transfusion

The efficacy of exchange transfusion as adjunctive therapy for severe malaria is controversial. No sufficiently powered, randomized, controlled study has been reported, although anecdotal case reports in the literature indicate benefits in selected groups of patients with hyperparasitaemia and organ failure (49,50). A meta-analysis of eight studies comparing survival rates associated with exchange transfusion to survival rates with antimalarial chemotherapy alone did not show improved survival rates in the former groups of patients. There were significant problems with the comparability of treatment groups in the studies reviewed, with higher levels of parasitaemia and more severe malaria in the group who received transfusions (51). Recent studies suggest that the benefits associated with exchange transfusion result from replacing the rigid, non-deformable parasitised and unparasitised red cells with fresh blood, and not from reducing parasite load or removal of toxins or cytokines (52).

Requirements for exchange transfusion include the availability of pathogen-free compatible blood, facilities for adequate clinical monitoring, and a haemodynamically stable patient. Exchange transfusion may be considered in a patient who is seriously ill and the parasitaemia exceeds 15%. Exchange should still be considered with parasitaemia in the range of 5-15%, if there are other signs of poor prognosis. There is no consensus of the volume of blood to be exchanged for a given parasitaemia and the volumes have varied from 4 litres to 20 litres. Blood may be exchanged using a double-lumen catheter or alternatively via haemodialysis (4,11). Successful red blood cell exchange using a cell separator has been reported (52).

PREDICTORS OF MORTALITY

In a study conducted in a well-established intensive-care unit in South Africa, despite appropriate chemotherapy with quinine, and standard intensive-care support including inotropic agents, ventilatory support and haemodialysis where appropriate, mortality was 28.5% in a group of 28 patients (24 adults and 4 children). Pregnancy was a major cause of unfavourable outcome. ARDS was the most important cause of death. High Apache II scores, high arterial lactate, and negative base excess in the first 24 hours of admission correlated with mortality. Admission haemoglobin, platelet count, level of parasitaemia and level of Glasgow Coma Scores in the first 24 hours were shown not be predictors of mortality. These parameters may be useful in the assessment of disease severity and in patient triage for ICU admission (53).

REFERENCES

1.   Greenwood B, Mutabingwa T. Malaria in 2002. Nature 2002; 415: 670-672.

2.   Marsh K. Malaria disaster in Africa. Lancet 1998; 352: 924-925.

3.   Baird JK. Age-dependent characteristics of protection v. susceptibility to Plasmodium falciparum. Ann Trop Med Parasitol 1998; 92: 367-390.

4.   World Health Organization (2000). Severe falciparum malaria. Trans R Soc Trop Med Hyg 2000; 94 (Suppl 1): S1/1-S1/90.

5.   Trape JF, Pison G, Preziosi MP, et al. Impact of chloroqine resistance on malaria mortality. C R Acad Sci Paris Serie III 1998; 321:689-697.

6.   Durrheim DN, Frieremans S, Kruger P, et al. Confidential inquiry into malaria deaths. Bull WHO 1999; 77: 263-266.

7.   Miller LH, Baruch DI, Marsh K. The pathogenic basis of malaria. Nature 2002; 415: 673-679.

8.   Chotivanich K, Udomsangpetch R, Simpson JA, et al. Parasite multiplication potential and the severity of falciparum malaria. J Infect Dis 2000; 181:1206-1209.

9.   Greenwood B, Marsh K, Snow R. Why do some African children develop severe malaria? Parasitol Today, 1991; 7: 277-280.

10.   Day NP, Hien TT, Schollaardt T, et al. The prognostic and pathophysiologic role of pro- and anti-inflammatory cytokines in severe malaria. J Infect Dis 1999; 180: 1288-1297.

11.   White NJ. The treatment of malaria. New Engl J Med 1996; 335: 800-806.

12.   Field JW. Blood examination and prognosis in acute falciparum malaria. Trans R Soc Trop Med Hyg 1949; 43: 33-48.

13.   Phu NH, Day NPJ, Piep PT, et al. Intraleucocytic malaria pigment and prognosis in severe malaria. Trans R Soc Trop Med Hyg 1995; 89: 200-204.

14.   Beadle C, Long GW, Weiss WR, et al. Diagnosis of malaria by detection of Plasmodium falciparum HRP-2 antigen with a rapid dipstick antigen-capture assay. Lancet 1994; 343: 564-567.

15.   Isaäcson M, Frean JA. African malaria vector in European aircraft. Lancet 2001; 357: 235.

16.   White NJ, Looareesuwan S, Warrell DA, et al. Quinine loading dose in cerebral malaria. Am J Trop Med Hyg 1983; 32: 1-5.

17.   Viriyayudhakorn S, Thitiarchakul S, Nachaisit S, et al. Pharmacokinetics of quinine in obesity. Trans R Soc Trop Med Hyg 2000; 94: 425-428.

18.   Kremsner P, Winkler S, Brandts C, et al. Clindamycin in combination with chloroquine or quinine is an effective therapy for uncomplicated Plasmodium falciparum malaria in children from Gabon. J Infect Dis 1994; 169: 467-470.

19.   White NJ, Looareesuwan S, Warrell DA, et al. Quinine. Pharmacokinetics and toxicity in cerebral and uncomplicated falciparum malaria. Am J Med 1982; 73: 564-557.

20.   Silamut K, Molunto P, Ho M, et al. Alpha 1-acid glycoprotein (orosomucoid) and plasma protein binding of quinine in falciparum malaria. Br J Clin Pharmacol 1991; 32: 311-315.

21.   Looareesuwan S, Phillips RE, White NJ, et al. Quinine and severe falciparum malaria in late pregnancy. Lancet 1985; ii: 4-8.

22.   Touze J-E, Heno P, Fourcade L, et al. The effects of antimalarial drugs on ventricular repolarization. Am J Trop Med Hyg 2002; 67 :54-60.

23.   Miller KD, Greenberg AE, Campbell CC. Treatment of severe malaria in the United States with a continuous infusion of quinidine gluconate and exchange transfusion. New Engl J Med 1989; 321: 65-70.

24.   Hien TT. An overview of the clinical use of artemisinin and its derivatives in the treatment of falciparum malaria in Vietnam. Trans R Soc Trop Med Hyg 1994; 88 (Suppl 1): S7-S8.

25.   Looareesuwan S. Overview of clinical studies on artemisinin derivatives in Thailand. Trans R Soc Trop Med Hyg 1994; 88 (Suppl 1): S9-S11.

26.   White NJ, Olliaro P. Strategies for the prevention of antimalarial drug resistance: rationale for combination therapy for malaria. Parasitol Today 1996; 12: 399-401.

27.   Murphy SA, Mberu E, Muhia D, et al. The disposition of intramuscular artemether in children with cerebral malaria: a preliminary study. Trans Royal Soc Trop Med Hyg 1997; 91: 331-334.

28.   Hien TT, Phy NH, Mai NTH, et al. An open randomized comparison of intravenous and intramuscular artesunate in severe falciparum malaria. Trans R Soc Trop Med Hyg 1992; 84: 584-585.

29.   The Artemether-Quinine Meta-Analysis Study Group A meta-analysis using individual patient data of trials comparing artemether with quinine in the treatment of severe falciparum malaria. Trans R Soc Trop Med Hyg 2001; 95: 637-650.

30. Hien TT, Day NPJ, Phu NH, et al. A controlled trial of artemether or quinine in Vietnamese adults with severe falciparum malaria. New Engl J Med 1996; 335: 76-83.

31. Weatherall DJ, Abdalla S. The anaemia of falciparum malaria. Br Med Bull 1982; 38: 147-151.

32. Looareesuwan S, Warrell DA, White NJ. Do patients with cerebral malaria have cerebral oedema? A computer tomography study. Lancet 1983; i: 434-437.

33. Looareesuwan S, Wilairatana P, Kriohna S. Magetic resonance imaging of the brain in patients with cerebral malaria. Clin Infect Dis 1995; 21: 300-309.

34. Warrell DA, Looareesuwan S, Warrell MJ. Dexamethsaone proves deleterious in cerebral malaria. A double-blind trial in 100 comatose patients. New Eng J Med 1982; 306: 313-319.

35. Crawley J, Waruiru C, Mithwani S, et al. Phenobarbitone prophylaxis in childhood cerebral malaria: Final results of a randomized, controlled intervention study. Proc Multilateral Initiative on Malaria Conference, South Africa, 1999.

36. Gordeuk V, Thuma P, Brittenham G, et al. Effect of iron chelation therapy on recovery from deep coma in children with cerebral malaria. New Engl J Med 1992; 327: 1473-1477.

37. Day NPJ, Phu NH, Loc PP. Malaria and acute renal failure. J R Coll Physicians Lond 1997; 31: 146-148.

38. Tran TT, Phu NH, Vinh H. Acute renal failure in patients with severe falciparum malaria. Clin Infect Dis 1992; 15: 874-880.

39. James MFM. Pulmonary damage associated with falciparum malaria. A report of ten cases. Ann Trop Med Parsitol 1985; 79: 123-138.

40. Brody MG, Kiel FW, Sheehy TW, et al. Acute pulmonary oedema in falciparum malaria. New Engl J Med 1968; 279:732-737.

41. Day NP, Phy NH, Mai NT, et al. The pathophysiologic and prognostic significance of acidosis in severe adult malaria. Crit Care Med 2000; 28: 1833-1840.

42. Tran TH, Day NP, Ly VC. Blackwater fever in southern Vietnam: a prospective descriptive study of 50 cases. Clin Infect Dis 1996; 23: 1274-1281.

43. White NJ. Pathophysiology. Clinics in Tropical Medicine and Communicable Diseases 1986; 1: 55-90.

44. Tanios MA, Kogelman L, McGovern B, Hassoun PM. Acute respiratory distress syndrome complicating *Plasmodium vivax* malaria. Crit Care Med 2001; 29: 665-667.

45. Beg MA, Khan R, Baig SM, et al. Cerebral involvement in benign tertian malaria. Am J Trop Med Hyg 2002; 67: 230-232.

46. Whitworth J, Morgan D, Quigley M, et al. Effect of HIV-1 and increasing immunosuppression on malaria parasitaemia and clinical episodes in adults in rural Uganda: a cohort study. Lancet 2000; 356: 1051-1056.

47. French N, Gilks CF. Fresh from the field: some controversies in tropical medicine and hygiene. HIV and malaria, do they interact? Trans R Soc Trop Med Hyg 2000; 94: 233-237.

48.   Cohen C, Karstaedt A, Govender N. Increase in severe malaria in HIV-positive adults in South Africa. XIV International AIDS Conference, Barcelona, 2002.

49.   Burchard GD, Kröger J, Knobloch J, et al. Exchange blood transfusion in severe falciparum malaria: retrospective evaluation of 61 patients treated with, compared to 63 patients treated without, exchange transfusion. Trop Med Internat Health 1997; 2 :733-740.

50.   Phillips P, Nantel S, Benny WB. Exchange transfusion as an adjunct to the treatment of severe falciparum malaria: Case report and review. Rev Infect Dis 1990; 12: 1100-1108.

51.   Riddle MS, Jackson JL, Sanders JW. Exchange transfusion as an adjunct therapy in severe *Plasmodium falciparum* malaria: A meta-analysis. Clin Infect Dis 2002; 34: 1192-98.

52.   White NJ. What is the future of exchange transfusion in severe malaria? J Infect 1999; 39: 185-186.

53.   Blumberg L, Lee RP, Lipman J, et al. Predictors of mortality in severe malaria: A two-year experience in a non-endemic area. Anaesth Intens Care 1996; 24 : 217-223.

2

Severe Malaria: North American Perspective

Monica E. Parise and Linda S. Lewis
Centers for Disease Control and Prevention, Atlanta, USA, and Health Studies Consulting, Medford, OR, USA

BACKGROUND

Plasmodium falciparum is responsible for essentially all of 1 million deaths annually that occur worldwide due to malaria and is the only one of the four human malaria species (others include *P. vivax*, *P. ovale*, and *P. malariae)* that clearly causes severe malaria.[1] *P. falciparum* is endemic in Africa, the Middle East, Oceania, Southeast Asia and India, and Central and South America. Persons living in highly malarious areas may become partially immune to malaria; however, this immunity is not fully protective and may wane during pregnancy or with time after a person leaves the endemic area.

Widespread drug resistance has complicated the clinical management of *P. falciparum*. The resistance of *P. falciparum* to chloroquine has been confirmed in all areas with *P. falciparum* malaria except the Dominican Republic, Haiti, Central America west and north of the former Panama Canal Zone, Egypt, and some countries in the Middle East. In addition, resistance to sulfadoxine-pyrimethamine is widespread in the Amazon River basin area of South America, much of Southeast Asia, other parts of Asia, and, increasingly, in parts of eastern and southern Africa. Resistance to mefloquine has been confirmed on the borders of Thailand with Burma (Myanmar) and Cambodia, in the western provinces of Cambodia, and in the eastern states of Burma (Myanmar).[2, 3]

EPIDEMIOLOGY OF MALARIA IN U.S. TRAVELERS

Non-immune travelers who visit malaria-endemic areas are highly susceptible to severe disease if they become infected with *P. falciparum*. Most imported *P. falciparum* malaria among American travelers is acquired in sub-Saharan Africa. From 1985 through 2001, 5,015 cases of *P. falciparum* among U.S. civilians were reported to the Centers for Disease Control and Prevention. Of these, 4,310 (85.9%) were acquired in sub-Saharan Africa; 278 (5.5%) in Asia; 300 (6.0%) in the Caribbean and Central or South America; and 127 (2.5%) in other parts of the world.

Among 70 fatal malaria cases that occurred among U.S. civilians between 1985 through 2001, 66 (94%) were caused by *P. falciparum*, of which 47 (71%) were acquired in sub-Saharan Africa.[2] The mortality rate from *P. falciparum* in the United States has been estimated at 4.2% with an age-specific mortality rate increasing from 0% on persons under age 20 to 7.9% in those 50 years and older.[4] Factors associated with increased mortality among north American travelers include older age, failure to take antimalarial chemoprophylaxis, delay in seeking medical care, misdiagnosis, and inadequate therapy.[4, 5]

The successful management of *falciparum* malaria includes recognition of infection by health care providers, rapid and accurate laboratory testing and prompt initiation of effective treatment.

DEFINITION OF SEVERE MALARIA

WHO defines severe malaria as occurring when: a patient with asexual *P. falciparum* parasitemia and no other confirmed cause of their symptoms with one or more of the following clinical or lab features: prostration, impaired consciousness, respiratory distress (acidotic breathing), multiple convulsions, circulatory collapse, pulmonary edema (radiological), abnormal bleeding, jaundice, hemoglobinuria, severe anemia.[1]

PATHOGENESIS OF SEVERE MALARIA

Malaria sporozoites are transmitted to humans via the bite of infected female *Anopheles* mosquitoes. Sporozoites invade hepatocytes, undergo exo-erythrocytic development and release asexual forms (called merozoites) into the bloodstream on average 6 to 14 days later. These parasites then invade susceptible red blood cells, and undergo an erythrocytic development phase—once mature, are released from red blood cells and continue the cycle.   Some of the asexual forms will differentiate into female and male gametocytes that are then capable of infecting a mosquito during a blood meal.   *P. falciparum* causes the preponderance of malaria mortality because, unlike the other malaria species, it invades red blood cells of all ages, leading to a high proportion of red blood cells infected and destroyed by the parasite and is the only malaria species that sequesters in the deep capillary beds, causing microvascular disease—both of which contribute to severe and compolicated malaria. The hemolysis, which may lead to severe anemia, hemoglobinuria, acute tubular necrosis, and acute renal failure.[1]

While *P. falciparum* ring forms are seen in circulating erythrocytes, later stages of the falciparum parasite are found sequestered in internal organs. Sequestration is thought to be caused by cytoadherence, rosette formation, and/or decreased deformability of the erythrocytes. Infected erythrocytes have a propensity for sticking to vascular endothelium through a specific molecular interaction between parasite adhesins (that are located on or under the surface of infected red blood cells) and ligands on endothelial cells.[1] Binding of uninfected erythrocytes to the surface of erythrocytes infected with mature forms of *P. falciparum* (called rosetting) may contribute to venular obstruction.[6]

Because parasite sequestration occurs in deep vascular beds of internal organs during severe falciparum malaria, clinicians should anticipate multiple organ involvement with corresponding symptoms and signs—for example, neurologic (impaired consciousness, seizures); respiratory (pulmonary edema); hematologic (abnormal bleeding, hemolysis with subsequent jaundice and hemoglobinuria); and renal (insufficiency).[1] In patients with cerebral malaria, sequestration was more evident in the brain than in other organs.[7] While sequestration is the essential pathological feature of severe falciparum malaria, the exact pathogenesis of how this phenomenon leads to cerebral malaria and the other complications of malaria is still unclear.   Proposed mechanisms include the mechanical obstruction of blood flow, systemic cytokine production, and local production and deposition of cytokines.[1, 8]

CLINICAL MANIFESTATIONS

The usual incubation period for *P. falciparum* is 12-14 days, although symptoms may occur as early as one week after exposure. The incubation period may be prolonged in persons taking chemoprophylaxis or who have received partial treatment. Severe malaria typically develops after 3-7 days of non-specific symptoms in adults,[1] but progression to death can occur as rapidly as 24 hours after onset of symptoms.[9] In addition, non-immune patients may have no signs of severe disease at presentation but can subsequently rapidly deteriorate even when on appropriate therapy.[10]

The presenting signs and symptoms of falciparum malaria are non-specific and may mimic a number of common ailments including gastroenteritis, pyelonephritis, pharyngitis, upper respiratory tract infection, and undifferentiated viral syndromes.[11] Among US citizens infected with falciparum malaria, presenting symptoms included fever and chills (88%), malaise and weakness (50%), gastrointestinal symptoms (nausea, vomiting, diarrhea) 43%, neurological symptoms (dizziness, confusion, disorientation, coma) 37%, myalgias (24%), headache (19%), and shortness of breath (9%).[5] Among persons living in endemic areas, children with severe malaria are more likely than adults to have impaired consciousness, acidotic breathing, multiple convulsions, and severe anemia.[1] Children are more likely to present with vomiting, hypoglycemia, and hyperpyrexia than adults.[12] In contrast, pulmonary edema and acute renal failure are more common in adults, with pregnant women at particular risk for pulmonary edema.[1]

Both adults and children may develop cerebral malaria, which often begins dramatically with a convulsion followed by persisting unconsciousness. The strict definition of cerebral malaria requires the presence of unarousable coma in a patient who has a blood smear demonstrating *P. falciparum* and no other identifiable cause. However, it is recommended that patients with malaria who have any degree of impaired consciousness or evidence of neurological dysfunction be treated aggressively. The Glasgow coma scale in adults and Blantyre coma scale in children is useful for determining the level of impaired consciousness.[1] Other clinical manifestations that may be present in persons with cerebral malaria include retinal hemorrhages, decerebration (hypertonicity, posturing, or opisthotonus), disorders of conjugate gaze, forcible jaw closure and teeth grinding (bruxism), and (uncommonly) absent corneal reflexes.[1, 13]

DIAGNOSIS

The clinical diagnosis of malaria can be challenging, given that there are no signs or symptoms that are pathognomic for malaria.[1, 14] The presence of rash, lymphadenopathy, or any sign of focal infection in a patient with suspected malaria should suggest a different or additional diagnosis. The diagnosis of malaria may be missed in patients whose primary presentation is acute renal failure, a bleeding diathesis, severe hemolysis, pulmonary edema or shock. In addition, heat stroke should be considered in the differential diagnosis of severe fever and altered consciousness, especially in unacclimatized visitors to the tropics or in those with a history of excessive exercise. Some persons with malaria develop diarrhea, which may be mistaken for infectious gastroenteritis.[1] In patients with altered consciousness, lumbar puncture should be performed to rule out bacterial meningitis.[15] During pregnancy and the puerperium, severe malaria must be distinguished from sepsis arising from infection in the uterus, urinary tract or breast. Cerebral malaria can be distinguished from eclampsia, as other features such as hypertension and edema are present with eclampsia but not with malaria.[1]

Malaria must be considered in the differential diagnosis of fever in the returned traveler but is often overlooked in non-malarious areas because the travel history is not investigated. Severe malaria may be mistaken for influenza, hepatitis, encephalitis, enteric fever, or psychosis, among other conditions. National U.S. surveillance data from 1959 through 1987 found that among US citizens infected with falciparum malaria, 37-40% were not diagnosed with malaria during the initial contact with the physician.[4, 5] In a review of *P. falciparum* cases presenting to a large emergency department in Los Angeles, a diagnosis of malaria was considered in only 60%; speciation of falciparum malaria was made in only 10% of the cases.[16] Similar findings have been reported by Kain and colleagues from Canada, where the diagnosis of malaria was missed at initial presentation in 61% of cases and 16% of patients reported seeing 3 or more physicians before the diagnosis of malaria was suspected. Thirty percent of patients with *P falciparum* in this Canadian series received inappropriate treatment.[17] In addition, patients themselves often delay seeking prompt medical attention, with only 54% of persons infected with *P. falciparum* seeking care within 6 days after the onset of illness.[4]

The diagnosis of malaria is confirmed by the identification of parasites in a Giemsa-stained blood film. While a Wright-Giemsa stain may be used, a Wright's stain alone will not reliably stain malaria parasites. Non-immune patients can have significant symptoms early in their course when the parasitemia levels are still so low that detection on a blood smear is difficult.[11] Thick blood smears are much more sensitive for the detection of low density parasitemia since 10 times the amount of blood can be examined compared to that examined on a thin film. However, because the red blood cells are lysed, persons inexperienced with the slide diagnosis of malaria may have more difficulty reading thick compared to thin smears.

In non-immune persons with no prior exposure to malaria, initial malaria symptoms may occur even with very low parasite densities. Consequently, serial blood smears should be examined by a person experienced in slide diagnosis of malaria.[11] Smears should be repeated every 6 to 12 hours for a minimum of 3 days to rule out malaria unless an alternative diagnosis becomes clear.[1, 11] Although there is a correlation between parasite density and disease severity, a patient can have severe disease without a high level of peripheral parasitemia because a large number of parasites are sequestered in the deep capillary beds.[1] Every febrile illness in an individual who has been in a malaria endemic area during the preceding 3 months and even up to a year, regardless of reported use of malaria prophylaxis use, requires a malaria smear.[11] Because *P.falciparum* malaria is a medical emergency, it is imperative that clinicians and laboratories process malaria blood smears on an urgent basis. Even in case of suspected uncomplicated malaria, blood smears should be collected and read on the same day that the patient presents.

If malaria is suspected but initial smears are negative, the follow-up smears should be done, as noted above, but other potential causes of fever should be pursued. Negative blood smears have been reported very rarely from patients with severe malaria[1, 18] and are hypothesized to be due to prior antimalarial drug treatment, a highly synchronized infection, or a high degree of sequestration. However, some of the reported cases have been poorly documented[18-20] and most likely have alternative diagnoses.

TREATMENT

Patients with severe malaria should be treated with a loading dose of parenteral antimalarial drugs to assure therapeutic blood levels as quickly as possible; most deaths from severe malaria occur within the first 24-48 hours. Oral antimalarial drugs (such as oral quinine, chloroquine, or mefloquine) are not recommended for the treatment of severe malaria as their absorption may be impaired in the severely ill patient. If severe malaria is strongly suspected but the first blood smear does not demonstrate parasites, a trial of antimalarial drugs should be given. If there is clinical evidence of severe malaria but the blood smear is reported as *P. vivax, P. ovale* or *P. malariae*, the patient should be treated for *falciparum* malaria in case of a mixed infection or misdiagnosis.[1] The treatment for falciparum malaria covers the other species.

Since 1991, quinidine gluconate has been the only parenterally administered antimalarial drug available in the United States.[21] The artemisinin derivatives are very efficacious and are used in some other countries for the treatment of severe malaria but are not available in the United States. The CDC is attempting to find a mechanism to make a parenteral artemisinin derivative available for cases of failure to respond to quinidine or severe quinidine intolerance.

Quinidine Gluconate

Quinidine gluconate is recommended for patients infected with *P. falciparum* who are unable to take oral medications, have high-density parasitemia (>5% of red blood cells infected), or have end organ complications such as cerebral malaria or acute renal failure.[14, 22] The recommended dosage is 10mg/kg salt (equivalent to 6.25 mg base/kg) infused intravenously over 1-2 hours followed by a continuous infusion of 0.02 mg salt /kg/min (0.0125 mg/kg/min quinidine base). An alternative regimen is an intravenous loading dose of 24 mg/kg quinidine salt (15mg/kg quinidine base) infused over 4 hours, followed by a 12mg/kg salt (7.5mg/kg base) infused over 4 hours every 8 hours, starting 8 hours after the loading dose.[23] Quinidine levels should be maintained in the range of 3-8 mg/L.[14, 15] At least 24 hours of quinidine infusion are recommended (or 3 intermittent doses).[1] Once the parasite density is < 1% and the patient can take oral medication, they can complete their treatment course with oral quinine at a dosage of 10 mg salt/kg every 8 hours. The treatment course of the quinolone derivative (intravenous quinidine followed by oral quinine) may be 3 days in areas without multidrug resistance but should be extended to 7 days in areas with and multidrug resistance.[14]

A second drug is generally used in combination with quinine. In areas without multidrug resistant malaria, using a second drug allows one to shorten the course of quinine to 3 days (otherwise, if quinine is used alone, a full course of 7 days of quinidine/quinine is recommended). In areas with multidrug resistance, such as in Southeast Asia where a decrease in sensitivity to quinine has been documented,[24] a second drug is required. Oral tetracycline (250mg every 6 hours), doxycycline (100 mg every 12 hours), or clindamycin (5 mg/kg base orally every 6 hours) for 7 days are options. There should be an overlap of 2 to 3 days for quinine/quinidine and tetracycline.[14] Patients unable to tolerate oral therapy can be given intravenous doxycycline hyclate 100mg every 12 hours for 7 days.

Pregnant women should receive treatment with quinidine/quinine as for non-pregnant patients.[1] Although there have been concerns voiced about adverse effects of quinine/quinidine on the fetus (congenital abnormalities) and on the pregnant uterus (inducing labor), no reports linking the use of quinidine with congenital defects have been identified and many of the reports of malformations with quinine have occurred at very high (abortifacent) doses.[25] Most importantly, this therapy is potentially life-saving for both mother and fetus. Studies in Thailand have not demonstrated an oxytocic effect at therapeutic doses.[26, 27] In addition to quinidine, pregnant women or children less than 8 years old can be treated with clindamycin.[14, 28]

Initial (including loading) doses of parenteral quinine or quinidine need not be reduced in persons with renal failure. If renal failure persists or the patient does not have improvement in their clinical condition, the maintenance dosage should be reduced by one-third to one half on the third treatment day[1] because the pharmacokinetic properties of the cinchona alkaloids ar altrered in malaria, with a contraction in the volume of distribution and reduction in clearance that is proportional to the severity of disease.[29]

Parenteral quinidine is more cardiotoxic than quinine and should be administered in an intensive care setting with continuous EKG and frequent blood pressure monitoring to avoid cardiotoxicity.[1, 22] At the dosages required for the treatment of falciparum malaria, quinidine may cause ventricular arrhythmia, hypotension, hypoglycemia, and prolongation of the QT interval.[30] The risk for serious ventricular arrhythmia is increased by bradycardia, hypokalemia, and hypomagnesemia.[23] The quinidine infusion should be slowed or stopped for an increase in the QRS complex by > 50%, a QT interval > 0.6 seconds, a QTc interval that is prolonged by more than 25% of the baseline value, or hypotension unresponsive to fluid challenge.[1, 14] Recent use of other drugs that may prolong the QT interval (e.g., quinine, halofantrine, and mefloquine) should be considered when determining whether a patient should receive a loading dose of quinidine gluconate.[30] Recommendations for administration of a loading dose of quinine, for which there is more experience to base decisions on as compared to quinidine, are to give it unless the patient has received more than 40 mg/kq quinine in the last 2 days or if they have received mefloquine in the last 12 hours.[1] Consulting a cardiologist and a physician with experience in treating malaria is advised when treating malaria patients in the United States with quinidine gluconate.[23] Glucose must be monitored closely as quinidine- (or quinine-) induced hyperinsulinemic hypoglycemia can occur, particularly in pregnant women.[26]

With the advent of newer anti-arrhythmic agents, quinidine gluconate has been removed from many hospital formularies and fewer clinicians have experience with the drug. To ensure the availability of quinidine gluconate in U.S. health care facilities, hospital drug services should consider maintaining or adding quinidine gluconate to formularies or alternatively, be able to immediately locate a nearby health care facility that stocks it. If a local source cannot be located, quinidine gluconate should be requested from the local or regional distributor. In the event that quinidine gluconate is needed acutely, pharmacists and clinicians should contact Eli Lilly Company directly, telephone (800) 821-0538 to arrange a rapid shipment of the drug. If further assistance is needed in obtaining quinidine gluconate or in managing patients with malaria, contact CDC's malaria hotline, (770) 488-7788 Monday-Friday 8am to 4:30pm EST or after hours, weekends and holidays, call CDC's security station at (404) 639-2888 and ask to have the on-call person for malaria questions paged.[30]

Other antimalarial drugs

Mefloquine is not recommended for use in the treatment of severe malaria because there is no parenteral preparation. Intravenous antibiotics such as doxycycline and clindamycin are too slow-acting to be used alone and must be used with quinidine or quinine. Intravenous chloroquine is approved but not marketed in the United States and would have limited usefulness since most *P. falciparum* imported into the United States is acquired in areas with high levels of chloroquine resistance. Although no parenteral preparation of atovaquone/progunail exists, recently, therapy with intravenous quinine followed by oral atovaquone/proguanil (Malarone™) was shown to be safe and efficacious in children with severe malaria in Kenya (personal communication C. Hedgeley).

ADJUNCTIVE THERAPY

Several ancillary therapies have been attempted in an effort to improve outcomes in severe malaria. A few of these have actually been shown to be detrimental, such as corticosteroids and heparin.[31, 32] Others have not shown benefit (pentoxiphylline, cyclosporin A, intravenous immunoglobulin, desferroxamine) or are unproven (deferiprone, osmotic/diuretic agents for cerebral malaria, dextran, prostacyclin, dichloroacetate for lactic acidosis).[1, 33, 34]

EXCHANGE TRANSFUSION

Exchange transfusion has been used in the treatment of severe malaria since 1974 with apparent benefit in some cases.[22, 35-37] However, there is no clear consensus on the indications, or on specifics such as volume to be exchanged. While it has not been proven beneficial in an adequately powered randomized controlled trial[1] and there have been case series of children in endemic areas with hyperparasitemia managed successfully without exchange transfusion,[38] as well as case reports[39] and series[40] of non-immune patients with severe malaria hyperparasitemia who have been cured without exchange transfusion, there is an increasing impression that exchange transfusion is beneficial in very sick patients.[1] A recent meta-analysis of 8 studies found no evidence for increased survival rate among patients receiving exchange transfusion but was limited in ability to draw conclusions given that, in the studies reviewed patients receiving exchange transfusion had higher levels of parasitemia and more severe malaria In a sub-group analysis, the authors attempted to control for parasite density and WHO criteria for severe malaria and still did not find a survival difference but that sub-analysis lacked the power to detect a significant difference in survival. These authors estimate, if one assumed a (large) reduction in mortality of 50%, one would need at least 400 participants for a definitive study on the benefit or lack thereof of exchange transfusion.[41]

CDC recommends that exchange transfusion be strongly considered for persons with a parasitemia level of more than 10% or if complications such as cerebral malaria, non-volume overload pulmonary edema, or renal compromise exist.[14] Others have suggested other criteria – for example, > 30% in the absence of clinical complications or > 10% in the presence of complications.[1] Exchange transfusion is thought to have beneficial effects by removing infected red cells, by improving the rheological properties of blood, and by reducing toxic factors such as parasite derived toxins, harmful metabolites, and cytokines.[42] The risks of exchange transfusion include fluid overload, febrile and allergic reactions, metabolic disturbances (e.g. hypocalcemia), red blood cell alloantibody sensitization, transmissible infection, and line sepsis and thus, the potential benefits of exchange transfusion should be weighed against the risks. The parasite density should be monitored every 12 hours until it falls below 1%, which usually requires the exchange of 8-10 units of blood in adults.[14]

The technical aspects of exchange transfusion have been discussed in an excellent review by Powell and Grima.[42] Since cell separators became available in the mid-1980s, the method of choice for exchange transfusion in the developed world has been automated exchange (as opposed to manual exchange). Automated exchange allows for exchange of a greater volume of blood in a substantially shorter period of time, with more effective reduction in parasitemia and less cardiovascular instability.[43] Since most automated cell separators remove only a single component, a RBC exchange (erythocytopheresis) is performed and the plasma, leukocytes, and platelets are returned to the patient—favourable outcomes have been documented with this procedure in several cases.[43-47] A reasonable exchange would be 1 volume red cell exchange or approximately 8 to 10 units of red blood cells in adults. Because of the theoretical advantage of also removing cytokines, some authors prefer to follow the RBC exchange with a 1-volume plasma exchange using fresh-frozen plasma as the replacement fluid. An alternative would be to use reconstituted whole blood for the automated exchange, obviating the need to perform 2 exchanges,[42] though this may be technically more complicated.

MANAGEMENT OF SEVERE MALARIA AND ITS COMPLICATIONS

The key principles for the successful management of severe falciparum malaria include: early suspicion of diagnosis; rapid clinical assessment; early treatment with drugs using optimal doses of an efficacious agent administered by the parenteral route; prevention or early detection an d treatment of complications (e.g. seizures, hypoglycemia); correct fluid, electrolyte and acid-base disturbances; avoidance of harmful ancillary measures (e.g. corticosteroids).

In addition to the immediate administration of efficacious parenteral antimalarial drugs, other important aspects of patient care in those with severe malaria should be assessed and managed. These include: 1) evaluate for hypoglycemia and treat if necessary 2) assess hydration status and administer fluids accordingly; 3) measure and monitor urine output, inserting a urinary catheter if necessary; 4) run serial laboratory tests, e.g. comprehensive metabolic panel, CBC, PT/PTT, serum lactate, arterial blood gases; 5) consider need for hemodynamic monitoring, such as indwelling arterial line, pulmonary artery catheter, and/or central venous pressure catheter; 6) treat high fever; 7) consider lumber puncture to exclude meningitis; and 8) evaluate and treat specific complications (see below).[1] The parasite count should be measured at least twice daily in all patients.[15] If the parasite count has not fallen at least 75% by 72 hours[48] after instituting antimalarial therapy, it should be rechecked. Upon repeat testing, if it is confirmed at the parasite count has not fallen appropriately, measures should be taken to investigate the problem (e.g. check quinidine level).

Fluid, electrolyte and acid base disturbances

Patients with severe malaria may present in various states of hydration. Dehydration and hypovolemia may contribute to hypotension, shock, and acute renal failure. However, fluid overload can precipitate non-cardiogenic pulmonary edema, especially in adults. Following rehydration, central venous pressure should be monitored and maintained at approximately 5 cm water (pulmonary-artery occlusion pressure < 15 mm Hg).[15]

Hyponatremia,[51] hypocalcemia, hypo- and hyperphosphatemia, hypo- and hypermagnesemia[49] have all been reported in patients with *P. falciparum* malaria. Electrolyte levels should be evaluated and alterations treated accordingly.

Metabolic acidosis occurs frequently in patients with severe malaria In most cases, even in patients with acute renal failure, the acidosis is attributable to lactic acidosis. Sequential studies have shown that lactate levels fall rapidly after treatment is begun, probably due to rehydration, cooling, and to the antiparasitic effects of the antimalarial drugs employed.[1] Alkali therapy has a limited role in the management of patients with lactic acidosis. Therapy with sodium bicarbonate has limited efficacy in raising the pH, and it produces an undesireable increase in the pCO_2 of the body fluids. Carbicarb (a commercially available buffer solution that is a 1:1 solution of sodium bicarbonate and disodium carbonate) does not produce the same increase in pCO_2 seen with sodium bicarbonate solutions.   Other buffers, such as the amine buffer tromethamine, or THAM, that do not generate pCO_2 are available, but clinical experience with it is much more limited than with bicarbonate. The administration of bicarbonate[50] should be considered if arterial pH falls below 7.10 and the patient is deteriorating. In this case, a trial infusion of bicarbonate can be attempted by administering one-half of the estimate bicarbonate deficit.   If cardiovascular improvement occurs, bicarbonate therapy can be continued. If bicarbonate therapy is followed by no improvement or further deterioration, further bicarbonate should not be given.[51] Arterial pH should be corrected slowly over 1-2 hours. Too rapid a correction can precipitate cardiac arrhthymias and paradoxical central nervous system acidosis.[1]

Dichloroacetate activates pyruvate kinase and has been shown to lower lactate levels without improving survival in patients with lactic acidosis.[52] Although it could potentially prove to be beneficial in treating the lactic acidosis associated with severe malaria,[53, 54] its efficacy is currently unproven.

Hypotension/shock

Possible etiologies for hypotension in severe malaria cases include hypovolemia, massive blood loss (gastrointestinal bleed or splenic rupture), pulmonary edema, and septicemia.[1]   Fluids should be administered, usually 0.9% saline initially followed by a blood product replacement or colloids. Bacterial infections are common in patients with severe malaria, and include pneumonia (especially in those patients comatose for more than 3 days), urinary tract infections in patient with indwelling catheters, intravenous line sepsis, and spontaneous septicemia, usually gram negative, presumably from the gastrointestinal tract.[15] If a patient's condition suddenly deteriorates, blood glucose should be checked to rule out hypoglycemia, cultures of blood (and urine, sputum, and catheter insertion sites, if indicated) performed, and empirical treatment for sepsis with broad-spectrum antibiotics should be initiated.[1, 15]

Pulmonary edema

Pulmonary edema may occur in adults but is rare in children. It frequently has its onset after 1-2 days of treatment and at times occurs when the patient seemed to be improving. It may be due either to fluid overload or increased capillary permeability as part of theadult respiratory distress syndrome (ARDS).[1]   Pulmonary artery occlusion pressure can be measured to differentiate fluid overload from ARDS. Volume overload should be treated with diuretics, careful fluid management, and oxygen. For volume overload or ARDS, mechanical ventilation with positive end expiratory pressure/ continuous positive airway pressure may be needed.

Renal Failure

Acute renal failure in severe falciparum malaria results from acute tubular necrosis. Tubular necrosis may be seen in persons experiencing multisystem dysfunction in the acute phase of *P. falciparum* malaria; in persons who have been successfully treated with antimalarial drugs and yet still develop renal failure, and in persons with massive hemoglobinuria.[1, 55] Malaria-associated renal failure is rare in children. Acute renal failure in malaria is most commonly oliguric or anuric but urine output may be normal or increased.[1]

The management of acute renal failure in severe malaria patients is the same as for any cause of acute renal failure,[55] with indications for hemodialysis: hyperkalemia, unresponsive metabolic acidosis, fluid overload, or uremic signs or symptoms. A rapidly rising serum creatinine (>2.5-3 mg/dl/day) was one of the most sensitive indicators of the need for dialysis. With dialysis, renal function can be expected to return after a median of 4 days, although some patients may requires dialysis for 2-3 weeks.[55, 56]

Cerebral malaria

Because hypoglycemia may be the underlying cause of impaired consciousness or convulsions, blood glucose should be checked. Vital signs should be monitored along with level of consciousness, Glasgow (or Blantyre) coma score.[1] A lumbar puncture should be performed to rule out bacterial meningitis.[1, 15] Cerebral edema is not thought to be part of the primary pathology of cerebral malaria, but may develop terminally or during prolonged intensive care if there is fluid overload or gross hypoalbuminemia. Deterioration in the level of consciousness and neurologic abnormalities in the absence of hypoglycemia are indications for computerized tomography of the head to rule out intracranial bleed, cerebral edema, or cerebral/ medullary herniation. Patients may need assisted ventilation in the event of central respiratory failure.[1] Most patients who survive cerebral malaria regain consciousness within 2-3 days, though it may occasionally take more than a week.[57]

Seizures

Generalized seizures occur in less than 20% of adults but more than 80% of infants with cerebral malaria.[15, 58] Seizures frequently mark the onset of coma or are followed by neurological deterioration.[1] Seizures may be focal and may be quite subtle in the unconscious patient.[15, 59, 60] Hyperpyrexia (fever >39 C) and hypoglycemia should be ruled out as a cause of seizures.[1] Seizures should initially be managed with diazepam (0.2 mg/kg at 5 mg/minute, which can be repeated in 5 minutes if needed) or lorazepam (0.1 mg/kg infused at 2 mg/min). Because of their short duration of action, these benzodiazepines should be followed by use of phenytoin (15-20 mg/kg at a maximum rate of 50 mg/minute). If the additional dose of phenytoin does not successfully control seizures, additional doses of 5 mg/kg may be administered up to a total cumulative dose of 30 mg/kg. For cases that do not respond to benzodiapepines and phenytoin, phenobarbital should be given at a maximum rate of 100 mg/min until seizures are controlled or till a maximum dose of 20 mg/kg is reached. Inhalational anesthesia and neuromuscular blockage may be needed in a small percentage of refractory cases; urgent neurological consultation should be obtained.[51, 61]

The use of prophylactic anticonvulsants has been considered for prevention of seizures in patients with cerebral malaria. Use of a single dose of phenobarbitone administered to children or adults prior to the onset of seizures has been suggested to prevent seizures from occurring.[62] However, given that a recent study found an increased mortality among children treated preventively with phenobarbitone,[63] the role of prophylactic anticonvulsants is unclear at this time.[64]

Hypoglycemia

Hypoglycemia occurs in approximately 8% of adults[65] and one-third of children[13, 66] with severe malaria. Pregnant women and children are especially susceptible to hypoglycemia. After rehydration, 5% or 10% dextrose should be administered by intravenous infusion and blood glucose should be monitored frequently. Further complicating blood glucose levels is the fact that quinidine can cause hyperinsulinemic hypoglycemia, usually commencing at least 24 hours after initiation of treatment.[15]

Fever

Although there is some evidence that fever may inhibit parasite growth,[67, 68] there are clearly negative effects of fever, such as febrile seizures in young children and increased metabolic demands, and currently there are insufficient data to recommend a change in the practice of fever management during malarial illness.[69] Hyperpyrexia should be treated with antipyretic drugs,[69] especially if rectal temperature is greater than 39 C.[1] Acetominophen can be administered; however, non-steroidal anti-inflammatory agents are not recommended given that thrombocytopenia is common and coagulation abnormalities may occur.

Anemia and bleeding disorders

Anemia can develop rapidly secondary to severe hemolysis, increased RBC destruction (phagocytosis of RBCs, hypersplenism), and decreased RBC production.[70] There is no specific hemoglobin or hematocrit level below whicha transfusion should be administered and the indications for red cell transfusion include evidence of impaired tissue oxygenation or ongoing coronary or cerebrovascular ischemia in patients with an adequate blood volume. In addition, correction of hemoglobin < 7 g/dL in patients with a history of active coronary artery disease, cerebrovascular insufficiency, or significant cardiac dysfunction is recommended.[51] Severe anemia may be best treated with exchange transfusion in patients with circulatory overload. Repeated transfusion may be required to offset abnormally rapid hemolysis of transfused red blood cells.[1]

Thrombocytopenia is often present. Laboratory evidence of activated coagulation is more commonly seen than is disseminated intravascular coagulation (DIC) with bleeding.[71, 72] However, DIC has been reported in series of patients in both endemic and non-endemic areas and is associated with other manifestations of severe illness such as renal, pulmonary, or cerebral end organ complications.[1, 73, 74] For this reason, drugs which increase the risk of bleeding (aspirin, corticosteroids, NSAIDs, heparin) should be avoided in patients with severe malaria.

Hemoglobinuria/ Blackwater fever

Hemoglobinuria or blackwater fever may develop in patients with severe malaria. Historically it was attributed to sensitization of RBCs to quinine after the intermittent use of this drug as a prophylactic. Hemoglobinuria in patients with malaria is seen under 3 scenarios: (a) in persons with G6PD deficiency who have received oxidant drugs or foodstuffs; (b) in persons with G6PD deficiency who have acute malaria and have recievd treatment with antimalarial drugs (hemolysis may either be due to the infection or the drugs); or (c) in persons with normal concentrations of G6PD but who have acute, often severe, malaria.[1] Despite the hypothetical relationship between antimalarial drugs and hemolysis, antimalarial drugs should be continued until parasitemia is resolved.[1]

Because quinidine is the only parenteral antimalarial drug available in the United States, the decision of when the patient can be changed to a second drug will need to be individualized and based on the patient's clinical condition.

PROGNOSIS

Death from malaria is preventable provided there is prompt attention given to the following 3 events--medical attention must be sought; an accurate diagnosis must be made; and efficacious treatment must be initiated. Unfortunately, there are too often delays in one or more of these components.

Several parameters have been examined for prognostic value in severe malaria. The following clinical features are associated with a poor prognosis in patients with severe malaria: impaired consciousness; repeated seizures (>3 in 24 hours); respiratory distress; substantial bleeding; and shock. Laboratory features associated with a poor prognosis are: serum creatinine >3 mg/dl; acidosis (plasma bicarbonate <15mmol/l); jaundice (serum total bilirubin >2.5mg/dl); hyperlactatemia (venous lactate > 45mg/dl); hypoglycemia (blood glucose<40 mg/dl); elevated aminotransferase levels (>3 times normal); parasitemia >500,000 parasites/mm^3 or >10,000 mature trophozoites and schizonts / mm^3; and 5% or greater neutrophils containing malaria pigment.[15]

REFERENCES

1. WHO, Severe falciparum malaria. World Health Organization, Communicable Diseases Cluster. Trans R Soc Trop Med Hyg, 2000. **94 Suppl 1**: p. S1-90.
2. Health information for the international traveler 2003-2004. Centers for Disease Control and Prevention. 2003, Atlanta: US Department of Health and Human Services, Public Health Service.
3. Singhasivanon, P., Mekong malaria. Malaria, multi-drug resistance and economic development in the greater Mekong subregion of Southeast Asia. Southeast Asian J Trop Med Public Health, 1999. **30 Suppl 4**: p. i-iv, 1-101.
4. Lobel, H.O., C.C. Campbell, and J.M. Roberts, Fatal malaria in US civilians. Lancet, 1985. 1(8433): p. 873.

5.   Greenberg, A.E. and H.O. Lobel, Mortality from Plasmodium falciparum malaria in travelers from the United States, 1959 to 1987. Ann Intern Med, 1990. **113**(4): p. 326-7.

6.   Rowe, J.A., et al., Short report: Positive correlation between rosetting and parasitemia in Plasmodium falciparum clinical isolates. Am J Trop Med Hyg, 2002. **66**(5): p. 458-60.

7.   MacPherson, G.G., et al., Human cerebral malaria. A quantitative ultrastructural analysis of parasitized erythrocyte sequestration. Am J Pathol, 1985. **119**(3): p. 385-401.

8.   Miller, L.H., et al., The pathogenic basis of malaria. Nature, 2002. **415**(6872): p. 673-9.

9.   Greenwood, B.M., et al., Mortality and morbidity from malaria among children in a rural area of The Gambia, West Africa. Trans R Soc Trop Med Hyg, 1987. **81**(3): p. 478-86.

10.   Moore, D.A., et al., Assessing the severity of malaria. Bmj, 2003. **326**(7393): p. 808-9.

11.   Freedman, D.O., Imported malaria--here to stay. Am J Med, 1992. **93**(3): p. 239-42.

12.   Brabin, B.J. and Y. Ganley, Imported malaria in children in the UK. Arch Dis Child, 1997. **77**(1): p. 76-81.

13.   Molyneux, M.E., et al., Clinical features and prognostic indicators in paediatric cerebral malaria: a study of 131 comatose Malawian children. Q J Med, 1989. **71**(265): p. 441-59.

14.   Zucker, J. and C.C. Campbell, Malaria: Principles of Prevention and Treatment, in Parasitic Diseases. 1993. p. 547-567.

15.   White, N.J., The treatment of malaria. N Engl J Med, 1996. **335**(11): p. 800-6.

16.   Kyriacou, D.N., et al., Emergency department presentation and misdiagnosis of imported falciparum malaria. Ann Emerg Med, 1996. **27**(6): p. 696-9.

17.   Kain, K.C., et al., Imported malaria: prospective analysis of problems in diagnosis and management. Clin Infect Dis, 1998. **27**(1): p. 142-9.

18.   White, N.J., S. Krishna, and S. Looareesuwan, Encephalitis, not cerebral malaria, is likely cause of coma with negative blood smears. J Infect Dis, 1992. **166**(5): p. 1195-6.

19.   Chia, J.K., M.M. Nakata, and S. Co, Smear-negative cerebral malaria due to mefloquine-resistant Plasmodium falciparum acquired in the Amazon. J Infect Dis, 1992. **165**(3): p. 599-600.

20.   Zucker, J.R. and C.C. Campbell, Smear-negative cerebral malaria due to mefloquine-resistant Plasmodium falciparum acquired in the Amazon. J Infect Dis, 1992. **166**(6): p. 1458-9.

21.   Availability and use of parenteral quinidine gluconate for severe or complicated malaria. MMWR Morb Mortal Wkly Rep, 2000. **49**(50): p. 1138-40.

22.   Miller, K.D., A.E. Greenberg, and C.C. Campbell, Treatment of severe malaria in the United States with a continuous infusion of quinidine

gluconate and exchange transfusion. N Engl J Med, 1989. **321**(2): p. 65-70.

23. Product Information: Quinidine Gluconate Injection: Uses and Dosage. 2001, Medscape Inc.

24. Pukrittayakamee, S., et al., Quinine in severe falciparum malaria: evidence of declining efficacy in Thailand. Trans R Soc Trop Med Hyg, 1994. **88**(3): p. 324-7.

25. Briggs, G.G., R.K. Freeman, and S.J. Yaffe, Drugs in pregnancy and lactation: a reference guide to fetal and neonatal risk. 6th ed. 2002, Philadelphia: Williams & Wilkins.

26. Looareesuwan, S., et al., Quinine and severe falciparum malaria in late pregnancy. Lancet, 1985. **2**(8445): p. 4-8.

27. McGready, R., et al., Randomized comparison of quinine-clindamycin versus artesunate in the treatment of falciparum malaria in pregnancy. Trans R Soc Trop Med Hyg, 2001. **95**(6): p. 651-6.

28. Pukrittayakamee, S., et al., Therapeutic responses to quinine and clindamycin in multidrug-resistant falciparum malaria. Antimicrob Agents Chemother, 2000. **44**(9): p. 2395-8.

29. White, N.J., et al., Quinine pharmacokinetics and toxicity in cerebral and uncomplicated Falciparum malaria. Am J Med, 1982. **73**(4): p. 564-72.

30. Availability and Use of Parenteral Quinidine Gluconate for Severe or Complicated Malaria. MMWR Morb Mortal Wkly Rep, 2000. **49**(50): p. 1138-40.

31. Warrell, D.A., et al., Dexamethasone proves deleterious in cerebral malaria. A double-blind trial in 100 comatose patients. N Engl J Med, 1982. **306**(6): p. 313-9.

32. Hoffman, S.L., et al., High-dose dexamethasone in quinine-treated patients with cerebral malaria: a double-blind, placebo-controlled trial. J Infect Dis, 1988. **158**(2): p. 325-31.

33. White, N.J., Not much progress in treatment of cerebral malaria. Lancet, 1998. **352**(9128): p. 594-5.

34. Smith, H.J. and M. Meremikwu, Iron chelating agents for treating malaria. Cochrane Database Syst Rev, 2003(2): p. CD001474.

35. Looareesuwan, S., et al., Plasmodium falciparum hyperparasitaemia: use of exchange transfusion in seven patients and a review of the literature. Q J Med, 1990. **75**(277): p. 471-81.

36. Hoontrakoon, S. and Y. Suputtamongkol, Exchange transfusion as an adjunct to the treatment of severe falciparum malaria. Trop Med Int Health, 1998. **3**(2): p. 156-61.

37. Saddler, M., et al., Treatment of Severe Malaria by Exchange Transfusion. N Engl J Med, 1990: p. 58.

38. Mordmuller, B. and P.G. Kremsner, Hyperparasitemia and blood exchange transfusion for treatment of children with falciparum malaria. Clin Infect Dis, 1998. **26**(4): p. 850-2.

39. Marik, P.E., Severe falciparum malaria: survival without exchange transfusion. Am J Trop Med Hyg, 1989. **41**(6): p. 627-9.

40. Burchard, G.D., et al., Exchange blood transfusion in severe falciparum malaria: retrospective evaluation of 61 patients treated with, compared to 63 patients treated without, exchange transfusion. Trop Med Int Health, 1997. **2**(8): p. 733-40.

41. Riddle, M.S., et al., Exchange transfusion as an adjunct therapy in severe Plasmodium falciparum malaria: a meta-analysis. Clin Infect Dis, 2002. **34**(9): p. 1192-8.

42. Powell, V.I. and K. Grima, Exchange transfusion for malaria and babesia infection. Transfus Med Rev, 2002. **16**(3): p. 239-50.

43. Mainwaring, C.J., et al., Automated exchange transfusion for life-threatening plasmodium falciparum malaria--lessons relating to prophylaxis and treatment. J Infect, 1999. **39**(3): p. 231-3.

44. Macallan, D.C., et al., Red cell exchange, erythrocytapheresis, in the treatment of malaria with high parasitaemia in returning travellers. Trans R Soc Trop Med Hyg, 2000. **94**(4): p. 353-6.

45. Files, J.C., C.J. Case, and F.S. Morrison, Automated erythrocyte exchange in fulminant falciparum malaria. Ann Intern Med, 1984. **100**(3): p. 396.

46. Sighinolfi, L., et al., Treatment of cerebral malaria by erythrocyte exchange. Recenti Prog Med, 1990. **81**(12): p. 804-5.

47. Zhang, Y., et al., Erythrocytapheresis for Plasmodium falciparum infection complicated by cerebral malaria and hyperparasitemia. J Clin Apheresis, 2001. **16**(1): p. 15-8.

48. Assessment and monitoring of antimalarial drug efficacy for the treatment of uncomplicated falciparum malaria. 2003, WHO: Geneva.

49. Davis, T.M., et al., Calcium and phosphate metabolism in acute falciparum malaria. Clin Sci (Colch), 1991. **81**(3): p. 297-304.

50. Adrogue, H.J. and N.E. Madias, Management of life-threatening acid-base disorders. First of two parts. N Engl J Med, 1998. **338**(1): p. 26-34.

51. Marino, P., The ICU Book. 2nd ed, ed. S.R. Zinner. 1998, Baltimore: Williams & Wilkins.

52. Stacpoole, P.W., et al., A controlled clinical trial of dichloroacetate for treatment of lactic acidosis in adults. The Dichloroacetate-Lactic Acidosis Study Group. N Engl J Med, 1992. **327**(22): p. 1564-9.

53. Krishna, S., et al., Dichloroacetate for lactic acidosis in severe malaria: a pharmacokinetic and pharmacodynamic assessment. Metabolism, 1994. **43**(8): p. 974-81.

54. Agbenyega, T., et al., Glucose and lactate kinetics in children with severe malaria. J Clin Endocrinol Metab, 2000. **85**(4): p. 1569-76.

55. White, N., Controversies in the management of severe falciparum malaria. Balliere's Clinical Infectious Diseases, 1995. **2**: p. 309-330.

56. Trang, T.T., et al., Acute renal failure in patients with severe falciparum malaria. Clin Infect Dis, 1992. **15**(5): p. 874-80.

57. Newton, C.R., T.T. Hien, and N. White, Cerebral malaria. J Neurol Neurosurg Psychiatry, 2000. **69**(4): p. 433-41.

58. Warrell, D.A., Cerebral malaria: clinical features, pathophysiology and treatment. Ann Trop Med Parasitol, 1997. **91**(7): p. 875-84.

59. Crawley, J., et al., Seizures and status epilepticus in childhood cerebral malaria. Qjm, 1996. **89**(8): p. 591-7.

60. Crawley, J., et al., Electroencephalographic and clinical features of cerebral malaria. Arch Dis Child, 2001. **84**(3): p. 247-53.

61. Treatment of convulsive status epilepticus. Recommendations of the Epilepsy Foundation of America's Working Group on Status Epilepticus. Jama, 1993. **270**(7): p. 854-9.

62. White, N.J., et al., Single dose phenobarbitone prevents convulsions in cerebral malaria. Lancet, 1988. **2**(8602): p. 64-6.

63. Crawley, J., et al., Effect of phenobarbital on seizure frequency and mortality in childhood cerebral malaria: a randomised, controlled intervention study. Lancet, 2000. **355**(9205): p. 701-6.

64. Meremikwu, M. and A.G. Marson, Routine anticonvulsants for treating cerebral malaria. Cochrane Database Syst Rev, 2002(2): p. CD002152.

65. White, N.J., et al., Severe hypoglycemia and hyperinsulinemia in falciparum malaria. N Engl J Med, 1983. **309**(2): p. 61-6.

66. White, N.J., et al., Hypoglycaemia in African children with severe malaria. Lancet, 1987. **1**(8535): p. 708-11.

67. Kwiatkowski, D., Febrile temperatures can synchronize the growth of Plasmodium falciparum in vitro. J Exp Med, 1989. **169**(1): p. 357-61.

68. Kwiatkowski, D., Tumour necrosis factor, fever and fatality in falciparum malaria. Immunol Lett, 1990. **25**(1-3): p. 213-6.

69. Meremikwu, M., K. Logan, and P. Garner, Antipyretic measures for treating fever in malaria. Cochrane Database Syst Rev, 2000(2): p. CD002151.

70. Menendez, C., A.F. Fleming, and P.L. Alonso, Malaria-related anaemia. Parasitology Today, 2000. **16**(11): p. 469-76.

71. Phillips, R.E., et al., The importance of anaemia in cerebral and uncomplicated falciparum malaria: role of complications, dyserythropoiesis and iron sequestration. Q J Med, 1986. **58**(227): p. 305-23.

72. Butler, T., et al., Blood coagulation studies in Plasmodium falciparum malaria. Am J Med Sci, 1973. **265**(1): p. 63-7.

73. Stone, W.J., J.E. Hanchett, and J.H. Knepshield, Acute renal insufficiency due to falciparum malaria. Review of 42 cases. Arch Intern Med, 1972. **129**(4): p. 620-8.

74. Punyagupta, S., et al., Acute pulmonary insufficiency in falciparum malaria: summary of 12 cases with evidence of disseminated intravascular coagulation. Am J Trop Med Hyg, 1974. **23**(4): p. 551-9.

3

Viruses in the Intensive Care Unit (ICU)

Guy A. Richards, Gunter Schleicher, Mervyn Mer
Department of Medicine, University of the Witwatersrand, Johannesburg, South Africa

INTRODUCTION

Infectious diseases in the developing world ICU usually involve bacterial sepsis resulting from community-acquired pneumonia, pelvic inflammatory disease, ruptured abdominal viscus (traumatic or spontaneous), necrotising fasciitis, or more exotic infections such as malaria. Despite their importance, viruses are rarely considered, except during outbreaks of hemorrhagic fever in which case they have short-lived notoriety.

Important viral infections in Africa often differ from those found in the United States of America (USA). This chapter will focus on viral hemorrhagic fevers, influenza, varicella, viral hepatitis, cytomegalovirus, measles and the respiratory syncitial virus.

Viral hemorrhagic fevers

The viral hemorrhagic fevers are generally characterized by a marked propensity for person-to-person spread and high mortality rates. This places them in the highest biohazard category (class 4) and renders them liable to control by the state in countries that have the relevant bio-safety regulations. The viruses themselves are numerous (1,2) (Table 1) and this chapter will confine discussion to those found in sub-Saharan Africa and some of those seen in South America and India. It is noteworthy that no cases of Hanta virus have been reported in Africa and in particular the Hanta virus pulmonary syndrome (3).

The viruses known or considered to be associated with hemorrhagic fever fall into three groups with respect to the primary means of transmission and reservoir hosts (Table 1). However, clinical manifestations are similar. They are febrile illnesses with an abrupt onset and usually with a short incubation period.   Headache, myalgia, lumbar pain, nausea, vomiting and diarrhea are frequent.   Hematological and serological findings are leukopenia (or leukocytosis), thrombocytopenia and elevated transaminases (2,4,5).   Coagulation profiles become progressively more abnormal and overt hemorrhagic features such as epistaxis; gingival bleeding and melena supervene from day 4 onward.   In those most affected, multiple organ system failure and death ensue.   Mortality is high, particularly with the filoviruses where Marburg has a fatality rate of 30%, Ebola Zaire 81-88%, with Ebola Ivory Coast intermediate between these two (4).

Table 1. Virus classification (HF, hemorrhagic fever)

Family	Diseases	Transmission
Arenaviridae	South American HF (Junin)	Rodent urine
	Lassa fever	Rodent urine
Bunyaviridae	Rift valley fever	Mosquitos / ticks
	Crimean Congo HF (CCHF)	Ticks
	HF with renal syndrome	Rodent
	Hantavirus (HV)	Rodent
Filoviridae	Filovirus HF	Unknown vectors
	Ebola, Marburg	
Flaviviridae	Yellow fever	Mosquito
	Dengue HF	Mosquito
	Kyasanur forest disease	Tick
	Omsk HF	Tick/rodent

The primary features of established illness are related to endothelial damage, hemorrhage and shock (1). Whereas direct cytopathogenesis appears to be a major mechanism of injury there is not extensive necrosis of endothelial cells (6). Endothelial dysfunction is more likely to be due to cytokine release as part of the systemic inflammatory response syndrome. Hemorrhage may be related to disseminated intravascular coagulopathy (DIC), but the presence of hepatic damage may confuse the picture. DIC is a regular feature of Marburg and Crimean-Congo HF but less frequent with arena-virus infections (5,7). Shock occurs as a consequence of hypovolemia. Only limited observations have been made in patients in whom shock persists after volume resuscitation and these have been contradictory with both an increased and a decreased systemic vascular resistance reported in association with a reduced cardiac index (3,8-10).

Therapy is supportive and directed toward ensuring adequate oxygen delivery. Hemorrhage is managed by replacement of appropriate clotting factors, platelets and blood as required, with monitoring of cardiac output mandatory given the uncertainty as to the etiology of hypotension (10,11). Positive transfer of human antibodies has not been proven to be of benefit in filovirus infection but may be of value in CCHF, although there has been no controlled trial (5). It is of value in treatment of Junin virus, with a reduction of mortality to $1 - 2$ % from 15-30% if initiated within the first 8 days of illness (12) and is also possibly of value in Lassa virus infection (13).

Promising results have been obtained with intravenous ribavirin in CCHF in South Africa and oral ribavirin in Pakistan but the discontinuation of the intravenous preparation has prevented proper evaluation (5,14). It is of particular value in Lassa fever with a reduction in mortality from 55% to 5% if begun within 6 days of onset of fever (15). In addition, ribavarin has some benefit in Argentine HF caused by Junin virus and reduces mortality in Hantan virus, which causes hemorrhagic fever renal syndrome in Asia (16).

Nursing and infection control are critical. It is possible that universal precautions may be sufficient to afford protection, however where a worry exists that airborne transmission is possible (this has been documented with the Reston and Zaire strains of Ebola virus in monkeys (4)), high level barrier nursing may be preferable utilizing isolation, protective clothing plus Hepa-filtered respirations. Infection control extends to the transport of specimens and their examination in the laboratory, where procedures should be in place to manage these materials.

Influenza
Influenza is increasingly being recognized as a cause of significant morbidity and mortality in the community, particularly among pediatric patients and the elderly (17,18). These viruses are subdivided into subtypes, which include host of origin, geographic location of first isolation, strain number and year of isolation (19). The antigenic description is of the hemaglutinin (HA) and neuraminidase (NA) and is given parenthetically.

Since 1933 major antigenic shifts have occurred in 1957 when the H_2N_2 subtype replaced the H_1N_1 subtype, in 1968 when the Hong Kong H_3N_2 virus appeared, in 1977 when the H_1N_1 virus reappeared and most recently in 1997 when the H_5N_1 Avian virus appeared (20,21). An epidemic was aborted in the latter case by eradication of the domestic bird population. Pneumonia is the most common complication, which occurs in high-risk patients including those with comorbid illness such as cardiovascular or pulmonary disease, diabetes, renal failure, immunosuppression, the elderly, or residents of nursing homes. The pneumonia may be primary (of viral origin) or secondary (related to bacterial infection). Primary viral pneumonia is the most severe, although the least common of the pneumonic complications and may occur in patients that are otherwise normal (22). Whereas secondary bacterial pneumonia has been reported to be the most frequent cause of death in previous pandemics (23), this was not the case in the most recent outbreak in Hong Kong (24). Where secondary bacterial pneumonia occurs, the most common pathogens are S. *pneumoniae* (48%). S. *aureus* (19%) and *H. influenzae*. The incidence of S. *aureus* is significantly increased in influenza epidemic years (18).

Other complications that may result in ICU admission are rhabdomyolysis, encephalitis, transverse myelitis and less commonly Reyes syndrome. Management is supportive, though new antiviral agents may play a role, particularly if administered early. All currently available drugs should be started within 2 days of onset of symptoms to be effective (25). The practical effectiveness of drugs such as oseltamivir and rimantidine remains to be determined (26).

Respiratory syncitial virus (RSV)
RSV is a frequently encountered, potentially severe infection in childhood. Disease is less severe in adults but may be more severe in the elderly, and in those with comorbid disease or immunocompromise (27,28).

Presentation is non-specific with fever, myalgia, arthralgias, wheeze and non-purulent or bloody sputum. X-ray changes are also non-specific and not helpful in the etiologic diagnosis of pneumonia. In one study in South Africa, 1288 patients admitted to hospital with an acute lower respiratory tract infection were identified over a 15-month period (29). Of these, pneumonia was diagnosed in 62.2%, bronchiolitis in 20.6% and laryngotracheobronchitis in 8%. 48.9% and 13% had moderate or severe disease respectively, the latter requiring admission to ICU. RSV enzyme immunoassay was positive in 16.4% of cases in all groups of diagnoses. Viral culture performed in 162 of the cases (12.6%), grew RSV in 11.7%, adenovirus in 3.7%, parainfluenza in 2.5% and influenza B in 0.6%.

Diagnosis is made most frequently by rapid antigen detection, but is not a routinely performed investigation outside of research studies. Enzyme immunoassays have sensitivities of 71-100% and specificities of 74-90% (30).

Treatment of RSV is supportive although nebulized ribavirin has proven effective in infants (31). This agent is not readily available in developing countries and it would be impractical to recommend routine enzyme testing for children or adults admitted to the ICU.

Varicella pneumonia

Varicella pneumonia represents a severe complication of varicella and most frequently occurs in adults. Estimates as to the incidence vary, with the highest being 50% of all adult cases and the overall incidence in the region of 14% (32-34). Varicella pneumonia has been reported to carry an overall mortality of between 10 and 30%. However, where mechanical ventilation is necessary, mortality is as high as 50% (35-37). Risk factors include cigarette smoking, pregnancy, immunocompromise and male gender (38-40). Whereas chickenpox is primarily a disease of childhood and less than 5% of reported cases occur in adults, more than 75% of all deaths take place in this group. Recent evidence indicates that there is an upward shift in the age at which chicken pox is contracted and as a consequence it is possible that more critically ill patients with varicella may be seen (41).

Varicella pneumonia causes an interstitial pneumonia with severe impairment of gas exchange. Pathologically this manifests as a florid immune reaction characterized by an interstitial pneumonitis with mononuclear cell infiltrates, capillary endothelial cell destruction, intra-alveolar exudates and hemorrhage, septal wall invasion by mononuclear cells and inflammatory changes in the bronchioles (38). The pneumonitis appears to be due to the host response rather than to specific virally mediated tissue injury.

Whereas usual therapy involves support and acyclovir, the benefit of the latter is uncertain (39). It is possible that acyclovir may hasten improvement in those that are less ill and do not require ventilation (35,38) but despite the recognition of limited efficacy it is still widely recommended as early primary therapy (39,41). A study performed in our ICU indicates that corticosteroids may dramatically alter the course of the most severe disease and should be considered in addition to antiviral therapy along with appropriate supportive care in any previously well patient with life threatening varicella pneumonia (42).

Little is known about the incidence and clinical cause of varicella pneumonia in HIV (human immunodeficiency virus) infected individuals (43,44). Patients with HIV or AIDS (acquired immunodeficiency syndrome) who are hospitalized with chickenpox appear to be at high risk for developing varicella pneumonia, which manifests in a similar clinical fashion to that in immunocompetent individuals. In a recent review conducted in a regional infectious diseases hospital affiliated to our institution, 58% of the patients who were hospitalized with chickenpox developed pneumonia (45). This incidence is significantly greater than in any previously reported study in immunocompetent patients.

Immunocompromised patients with varicella pneumonia have previously been reported to do poorly, with mortality as high as 50%, despite prompt initiation of antiviral therapy and supportive care (45). Our experience suggests that response to adjunctive corticosteroid therapy in patients with HIV/AIDS is as favorable as in immunocompetent patients.

Interestingly, recurrent varicella pneumonia requiring acute treatment followed by secondary antiviral prophylaxis in an HIV-infected adult patient has been described, analogous to other AIDS complicating opportunistic infections (43).

Hepatitis
The proportion of viral hepatitis infections that progress to acute liver failure caused by viruses is very low, occurring in less than 1% of patients with acute A or B hepatitis. However, viruses account for between 30-60% of all cases of liver failure (46,47). Most of these are related to hepatitis B (HBV) and a relatively smaller proportion to A (HAV) or other newly identified viruses. Fulminant hepatic failure is defined either as acute liver disease, occurring in the absence of pre-existing liver disease, which leads to encephalopathy within 8 weeks of onset of symptoms, or as liver disease, which leads to encephalopathy within 2 weeks of onset of jaundice (47). Clinical features are often non-specific, such as nausea and vomiting with progression to encephalopathy and coma. The prognosis is inversely proportional to the degree of encephalopathy.

Hepatitis A is an RNA virus transmitted by the fecal-oral route. Hepatitis B is a DNA virus and accounts for 40% to 75% of virally caused fulminant hepatitis. Transmission is via sexual contact, transplacentally, parenterally and in particular occupationally. Hepatitis D is an incomplete RNA virus that requires the presence of hepatitis B virus in order to infect an individual. It is an important cause of fulminant hepatitis and aggressive chronic hepatitis in HBV carriers (48). Hepatitis E virus is an RNA virus transmitted by the fecal-oral route and possibly parenterally which for unknown reasons carries a high mortality from fulminant hepatitis in pregnant women and is also the commonest cause of fulminant hepatitis in India (49,50). Hepatitis C virus is an RNA virus of the flaviviridae family and is responsible for 20% of acute hepatitis, 70% of chronic hepatitis and 40% of end-stage cirrhosis in Europe. 80% of patients infected with hepatitis C develop chronic infection consisting of either chronic hepatitis, fibrosis or cirrhosis. It is also not usual for it to cause a fulminant hepatic failure (51) but this occurs only in areas with high hepatitis C serum prevalence (47).

The diagnosis of HAV is made on the basis of the detection of high levels of IgM antibodies in the serum. In fulminant hepatitic failure caused by HBV, the widespread hepatic necrosis that occurs as a consequence of immune mediated lysis of infected hepatocytes may result in IgM anti-hepatitis B core (anti-HBc) being the only marker of hepatitis B, as hepatitis B surface antigen and hepatitis B DNA may be absent from serum (46).

Other viruses in the herpes group (cytomegalovirus, herpes simplex and Ebstein Barr virus), adenovirus and influenza virus may rarely cause fulminant hepatic failure.

Treatment involves identification of the cause and if possible, specific therapy. If facilities are available, patients with grade 2 encephalopathy or greater should be transferred to a liver transplant center (52). Supportive therapy, involves hemodynamic management, ventilation, prevention and treatment of hemorrhage, dialysis, therapy of co-existent sepsis and electrolyte disturbance, and management of intracranial pressure (47). Orthoptic liver transplantation is not frequently available in developing countries, but in appropriate patients has been shown to improve survival significantly (52).

Measles
Measles is a frequently encountered disease in the ICU in developing countries. The presence of malnutrition and often the lack of an effective vaccination programme combine to convert this "harmless" childhood infection into a major killer. In one case series of 15 patients admitted to an ICU during a measles epidemic, 11 were malnourished and none had been vaccinated. All 15 required mechanical ventilation for pneumonia and ARDS, 4 died and 4 developed long-term sequelae, i.e. chronic lung

disease, subacute sclerosing panencephalitis, hemiplegia or partial amputation of a limb (53). Young adults are not exempt from the ravages of this disease. In a study from Greece, 424 previously healthy young males were hospitalized with measles. 41 had bacterial pneumonia on admission and 71 developed pneumonia in hospital or post discharge (54). In another study of 68 adult patients admitted with measles diagnosed on clinical and serological grounds, 9 required intensive care, six mechanical ventilation for approximately 15 days, and two deaths occurred. Prior vaccination history was not available (55).

It would be best to avoid measles entirely by means of vaccination, however once contracted a study conducted in South Africa indicated that Vitamin A supplements reduce morbidity and mortality significantly and concluded that these should be given regardless of the presence or absence of clinical evidence of Vitamin A deficiency (56).

Herpes simplex virus type 1 encephalitis

Herpes encephalitis is the most common cause of fatal sporadic encephalitis in the United States, accounting for 10-20% of the annual 20 000 cases of viral encephalitis. No accurate figures are available as to the incidence of this disease in the developing world, however we see sporadic cases in our ICU. This disease occurs in all age groups with the development of focal encephalitis with progressive oedema and necrosis. The syndrome is characterized by rapid onset of fever, seizures, focal neurological signs and impaired level of consciousness (57-59).

In adults the etiology is herpes simplex type 1 whereas in neonates type 2 may also be involved and confers a worse outcome. Brain biopsy is no longer a routine diagnostic test and polymerase chain reaction assays are considered the best non-invasive technique (60,61). This test is positive early in the disease and remains so during the first week. Early aggressive antiviral therapy with acyclovir improves mortality and reduces subsequent cognitive impairment. Acyclovir provides better outcome than vidaribine.

Cytomegalovirus (CMV)

Whereas CMV is usually asymptomatic, severe morbidity may occur in the premature neonate and organ and bone marrow transplant recipients (62). Seronegative patients receiving a seropositive organ transplant will develop a primary infection in 70-90% of cases (63,64). Seropositive patients will develop CMV infection by superinfection or reinfection in 60-100% of cases. Primary infection is the most likely type and is also usually more severe. Those who receive anti-thymocyte or anti-lymphocyte globulins and those who have bone marrow transplants also have more severe disease and a mortality of 75-90% (65,66). CMV infection occurs most frequently 3-16 weeks after transplantation. Manifestations include, fever, hepatitis, leukopenia and thrombocytopenia. The most important condition resulting in admission to ICU is interstitial pneumonitis. This is associated with variable changes in the chest radiograph, most commonly showing diffuse bilateral infiltrates, but focal consolidation or nodules may occur. In the developing world CMV is more frequently seen in association with HIV. Despite the clear association with mortality in organ transplant recipients, in particular bone marrow transplants, the significance of CMV as a pathogen in patients with AIDS is unclear. Autopsy studies demonstrate that although CMV pneumonitis is frequently present it is not commonly found as the sole pathogen (67,68). In addition bronchoalveolar lavage specimens are also positive for CMV in more than 60% of patients (69). CMV has also been reported to be a potential cause of ventilator associated pneumonia in immunocompetent patients (70) and it is suggested that CMV should be considered as a possibility in patients not responding to antimicrobials or if there is evidence of other hospital outbreaks of viral infection particularly in the pediatric wards (71).

Diagnosis is made most frequently with an antigenemic assay incorporating antibodies directed at the pp65 matrix protein of the CMV virus (72). This test has gained acceptance particularly in immunocompromised hosts and correlates with viremia (72,73). The polymerase chain reaction has an even higher sensitivity, but is not always widely available (74).

The mainstay of therapy for solid organ transplant is Ganciclovir, which appears to reduce morbidity (63, 75,76). In contrast, Ganciclovir is not effective in bone marrow transplant recipients. It should not therefore be used as a single agent therapy in these patients (77, 79). It is possible that combinations with immunoglobulin or cytomegalovirus immunoglobulin may be of value (80,81).

Severe Acute Respiratory Syndrome (SARS)
On November 16th 2002 an unusual respiratory illness was reported in Guangdong Province, Southern China, which was designated the severe acute respiratory syndrome (SARS)(82). Subsequent world -wide transmission was initiated by a doctor who traveled to Hong Kong, where he infected 10 guests in the same hotel (83).

A global alert was issued by the WHO on 12/3/03, an unprecedented step, which nevertheless was proven to be appropriate when 3 days later, as a consequence of this alert, similar cases were identified in Singapore and Canada.   Early international recognition of an impending crisis was precipitated in part by the detailed report by WHO clinician Carlo Urbani, who subsequently himself demised from SARS (84). Local spread of this disease occurred in Vietnam, Canada, Hong Kong, Singapore, China and Taiwan.

The Organism. Tissue culture isolation and electron microscopy resulted in rapid identification of the culprit virus as a novel coronavirus only distantly related to any that had previously infected humans (85,86). It is likely that it originated in animals, but it differs from all previously known coronaviruses in that most cause disease in only one host species whereas this virus appears also to have acquired the ability to infect humans.

The high concentration of viral DNA in the sputum suggested that droplet spread was the main mode of transmission. Lack of antibody in the general community indicated that this virus had not circulated widely in humans.

Diagnosis. The rapid sequencing of the genome   allowed early development of diagnostic tests. A number of PCR protocols have been developed (87).   These tests have high sensitivity, but a negative test cannot rule out infection. SARS follows an unusual pattern in that during the initial phases of the illness, virus shedding is relatively low. Because shedding peaks in respiratory specimens and stool only at around 10 days after onset of clinical illness, tests of very high sensitivity, which do not yet exist, are necessary (88).

Virus culture is extremely demanding and not useful for rapid diagnosis, however ELISA, immunofluorescence and neutralization tests will soon be available commercially. Detectable immune responses begin at day 5 or 6 but reliable antibody tests are available only after about day 10 following onset of symptoms. Seroconversion or a fourfold rise in titre indicates recent infection.

Diagnostic tests currently have severe limitations and extreme caution should be used before excluding the possibility of SARS on the basis of a test alone. Suspicious laboratory features include lymphopenia, thrombocytopenia and elevated lactate dehydrogenase levels (82).

Clinical Definition. From the perspective of clinicians, where local transmission has occurred, all cases of community-acquired pneumonia are suspect. Otherwise a travel history to an affected country or contact with an infected patient is essential. The WHO case definition as of 1/5/03 for a suspected or probable case of SARS is a useful resource (88).

Clinical Features. In a recent study by Peiris (89), all patients became apyrexial within 48 hours, but fever recurred in 85% (64 patients) at a mean of 8.9 days (± 3.1). In only 10 of these was nosocomial bacterial sepsis the cause. Between days 13 and 15, 23% of patients had another episode of fever. Radiological worsening occurred in 80% of patients at a mean of 7.4 (± 2.2 days), 61% subsequently improved, 15% had remained unchanged at the time of writing and 24% had progressed further to a diffuse ground glass appearance at a mean of 12.0 (± 4.4) days. 44% developed desaturation of less than 90% in room air at 9.1 (± 4.2) days after onset of symptoms. 32% (24) required ICU at a mean of 11.0 (± 6.4) days of whom 19 were intubated and 15 required mechanical ventilation for ARDS. The development of ARDS had a bimodal pattern with a peak at 11 days and another at 20 days. On univariate analysis, the risk factors for development of ARDS were age, male sex, chronic hepatitis B carriage, raised creatinine and recurrence of fever. IgG seroconversion had a 93% sensitivity at day 28 even with corticosteroid therapy; however, nasopharyngeal viral RNA detection was present in only 32% at presentation

In the study by Lee (82) 23.2% were admitted to ICU, all for respiratory failure. Mechanical ventilation was required in 19 (13.8%). 5 died (3.6%) all of whom had co-existing conditions. Multivariate analysis defined age [Odds Ratio (OR)/per decade of life 1.80 (1.16-2.81), p = 0.009; high peak LDH, O.R. per 100 units 2.09 (1.28-3.42) p = 0.003 and a neutrophil count that exceeded the upper limit of normal at presentation, O.R. 1.60 (1.03-2.50) p=0.04] as predictors of mortality. In those admitted to ICU, dramatic increases in lung opacity, shortness of breath and hypoxia occurred at a median of 6.5 days (range 3 to 12).

In a further study by Booth (90), 20% were admitted to ICU and of these 8 died i.e. there was a 6.5% 28-day mortality. Diabetes, relative risk (RR) 3.1 (95% CI 1.4-7.2) p=0.01, and other comorbid conditions, RR 2.5 (1.1-5.8) p=0.03, were independently associated with outcome.

Treatment. Viral amplification may be associated with cellular damage by cytolysis or immunopathological mechanisms (91). Once an immune response is mounted auto-immune tissue injury may occur. This has been the rationale for corticosteroid therapy in this condition (92).

Interestingly SARS behaves similarly to varicella, in that the disease is more severe in adults, pneumonic manifestations may occur some days after the onset of clinical symptoms and dramatic responses have been apparent after the use of corticosteroids (see varicella). Whereas there has been concern regarding the use of corticosteroids there have also been many proponents, particularly from those at the coalface in Hong Kong.

Rivabirin was initially the antiviral of choice since it is an effective treatment of fulminant hepatitis in mice infected with the mouse hepatitis coronavirus (93). However, no anti-viral has been reported to be clinically effective in humans. This drug is extremely expensive in its intravenous form and Health Canada recently stated that it would no longer provide access to ribavirin because of side effects and lack of clear efficacy (94).

So et al (91) have described a standard protocol for the management of SARS which involves the administration of a combination of ribavirin and corticosteroids for those patients with:
- extensive or bilateral chest radiographic involvement,
- persistent chest radiographic involvement and high fever for 2 days
- or worsening clinical, radiographic or laboratory findings, or oxygen saturation less than 95% in room air.

Late administration of corticosteroids appears to be less effective, correction of dose according to body weight results in more rapid improvement of symptoms in obese patients, and step-down within 2-3 days resulted in re-bound in some patients. Steroids in this protocol were administered in high doses beginning with methylprednisone (1mg/kg 8 mg x 5 days) and weaning over 21 days. There was no mortality and only 4 required short periods of non-invasive ventilation.

Mortality. The WHO has revised its initial estimates of the case fatality rates on the basis of more complete data from China, Hong Kong, Singapore and Vietnam (95). Mortality varies according to age; being less than 1% aged 24 years or younger, 6% aged 25-44, 15% aged 45-64 and greater than 50% aged 65 and older

Infection Control. SARS is highly contagious, specifically to health care workers and in particular where the index case is not immediately identified as having the disease. At the Prince of Wales Hospital, a patient was admitted on the 4/3/03 with "pneumonia" and was discharged well on the 11/3/03. On the 10/3/03 however, 18 health care workers became ill and a further 50 potential cases were identified on that day. By March 25th, 156 patients had been admitted to Prince of Wales Hospital with SARS, all traceable to this index patient (96). In Singapore similarly, initial transmission occurred from an index case to 12 patients, 10 of whom were primary contacts and 5 of whom were health care workers (97). In this latter case, Singaporean media labeled this patient a super-spreader, a concept that, although as yet inadequately defined may possibly be correct (98,99). Some patients do appear to spread the disease more readily and this may be related to the rate of shedding of viral particles. In Toronto amongst 144 cases 73 (51%) were health care workers and the subsequent outbreak of at least 90 patients occurred as a consequence of a cluster of unrecognized patients that had been admitted to North York Hospital in Toronto (90).

Infection control measures should include negative pressure wards, the use of N95 masks, gloves at all times, disposable impermeable gowns and eye protection. Hand washing after removal of gloves and avoidance of touching nose, eyes and mouth if at all possible are the most important practical measures (96).HCWS should be cohorted to decrease the number of people exposed and visiting should be strictly limited. Alcohol, phenol and quaternary ammonium based disinfectants are highly active against coronavirus. Certain features appear to enhance spread, in particular overcrowding of hospital wards, outdated ventilation systems and the use of nebulizers in the ward environment (96). Endotracheal intubation, open suctioning of respiratory secretions, the use of Bi-PAP and high frequency oscillatory ventilation in which air is forced around the face mask appear to be some of the most high- risk procedures

Numerous sources describing adequate infection control procedures are available, as well as those on the CDC website (99-101). The primary factor responsible for transmission seems to be inadequate training in or compliance with infection control procedures (102).

CONCLUSION

Whereas viruses are not usually considered to be important causes of ICU admission this review has demonstrated this perception to be incorrect.

Viruses and their manifestations differ from continent to continent and hemisphere to hemisphere and it is essential that the intensivist be familiar with diagnosis and management of these ubiquitous organisms.

REFERENCES

1.   Peters C J. Pathogenesis of viral hemorrhagic fevers. In: *Viral Pathogenesis.* Nathanson N, ed. Philadelphia: Lippincott-Raven, 1995.
2.   Swanepoel R. Viral hemorrhagic fevers in South Africa: History and national strategy. S Afr J Sci 1987, 83: 80-88.
3.   Duchin J S, Koster F, Peters C J, et al. Hantavirus pulmonary syndrome: A clinical description of 17 patients with a newly recognized disease. N Engl J Med 1994; 330: 949-955.
4.   McCormick JB, Fisher Hoch SP. Filoviruses. In: *Kass Handbook of Infectious Diseases. Exotic Viral Infections.* Porterfield JS, ed. London: Chapman and Hall, 1995.
5.   Swanepoel R. Crimean-Congo hemorrhagic fever. In: *Zoonoses: Biology, Clinical Practice and Public Health Control.* Palmers S, Lord Soulsby, Simpson D, eds. Oxford: Oxford University Press, 1998.
6   Feldman H, Bugary H, Mahner F, et al. Virus induced endothelial permeability triggered by affected macrophages. FASEB J 1994; 8: 156-272.
7.   Molinas FC, Bracco MM, Maiztequi JI. Hemostasis and the complement system in Argentine hemorrhagic fever. Rev Infect Dis 1989; 11 (Suppl 4): S762-S767.
8.   Entwisle C, Hale E. Hemodynamic alterations in hemorrhagic fevers. Circulation 1957; 15: 414-425.
9.   Qian C, Jarhling PB, Peters CJ, et al. Cardiovascular and pulmonary responses to Pichinde virus infection in strain 13 guinea pigs. Lab Anim Sci 1994; 44: 600-607.
10.   Van Eeden PJ, Van Eeden SF, Joubert R, et al. A nosocomial outbreak of Crimean-Congo hemorrhagic fever at Tyberberg Hospital (part II). Management of patients. S Afr Med J 1985; 68: 718-721.
11.   Richards GA, Murphy S, Jobson R, et al. Unexpected Ebola virus in a tertiary setting: Clinical and epidemiologic aspects. Crit Care Med 2000; 28: 240-244.
12.   Enria D, Briggiler A, Fernandez N, et al. Dose of neutralizing antibodies for Argentine hemorrhagic fever. Lancet 1984; 2: 255-256.
13.   Jahrling P, Frame J, Rhoderick J, Monson MH. Endemic Lassa fever in Liberia. IV. Selection of optimally effective plasma for treatment by passive immunization. Trans Roy Soc Trop Med Hyg 1985; 79: 380-384.
14.   Fisher-Hoch SP, Khan JA, Rehman S et al. Crimean-Congo hemorrhagic fever treated with oral ribavirin. Lancet 1995; 346: 472-473.
15.   McCormick JB, King IJ, Webb PA, et al. Lassa fever: effective therapy with ribavirin. N Engl J Med 1986; 314: 20-26.
16.   Huggins JW, Hsiang CM, Cosgriff TM, et al. Prospective, double blind, concurrent, placebo-controlled clinical trial of intravenous ribavirin therapy of hemorrhagic fever with renal syndrome. J Infect Dis 1991; 164: 119-1127.
17.   CDC update: influenza activity – United States 1999 – 2000 season. MMWR Morb Mortal Wkly Rep 2000; 49: 173-177.
18.   Barker WA, Mullooly JP. Pneumonia and influenza during epidemics. Arch Int Med 1992; 142: 85-89.

19. Lamb RA, Krug RM. Orthomyxoviridae: The viruses and their replication. In: *Fields Virology*, 3rd edition. Fields B N, Knipe D M, Howley P M, eds. Philadelphia: Lippincott-Raven, 1996.

20. To K, Chan P, Chan K, et al. Pathology of fatal human infection associated with avian influenza A H_5N_1 virus. J Med Virol 2001; 63: 242-246.

21. Yuen K, Chan P, Peiris M, et al. Clinical features and rapid viral diagnosis of human disease associated with avian influenza A H_5N_1 virus. Lancet 1998; 351: 467-471.

22. Lyytikainen O, Hoffman E, Timm H, et al. Influenza A outbreak amongst adolescents in a ski hostel. Eur J Clin Microbial Infect Dis 1998; 17: 128-130.

23. Kilbourne ED. Epidemiology of influenza. In: Kilbourne E D, (ed). *The influenza viruses and influenza*. London: Academic Press, 1995.

24. Schwarzman SW, Alder JL, Sullivan RF, et al. Bacterial pneumonia during the Hong Kong influenza epidemic of 1968-1969. Arch Intern Med 1971; 127: 1037-1041.

25. Couch RB. Influenza: prospects for control. Ann Intern Med 2000; 133: 992-998.

26. Oliveira E C, Marik P E, Colice G. Influenza pneumonia: A descriptive study. Chest 2001; 119: 1717-1723.

27. Harrington RD, Hooton TM, Hackman RC, et al. An outbreak of respiratory syncytial virus in a bone marrow transplant center. J Infect Dis 1992; 165: 987-993.

28. Akimoto CH, Cram DL, Root RK. Respiratory syncytial virus infections on an adult medical ward. Arch Intern Med 1991; 151: 706-708.

29. Hussey G, Apolles P, Ardense Z, et al. Respiratory syncytial virus infection in children hospitalized with acute lower respiratory tract infection. S Afr Med J 2000; 90: 509-12.

30. Kellog J. Culture vs direct antigen assays for detection of microbial pathogens from lower respiratory tract specimens suspected of containing the respiratory syncytial virus. Arch Pathol Lab Med 1991; 115: 451-458.

31. Smith DW, Frankel LR, Mathers LH, et al. A controlled trial of ribavirin in infants receiving mechanical ventilation for severe respiratory syncytial virus infection. N Engl J Med 1991; 325: 24-29.

32. Hockberger RS, Rothstein RJ. Varicella pneumonia in adults: Spectrum of disease. Ann Emerg Med 1986; 15: 931-934.

33. Mermelstein RH, Freireich AW. Varicella pneumonia. Arch Intern Med 1961; 55: 456-63.

34. Weber DM, Pellechia JA. Varicella pneumonia in adults: Spectrum of disease. Ann Emerg Med 1986; 15: 931-934.

35. Schlossberg D, Littman M. Varicella pneumonia. Arch Intern Med 1988; 148: 1613-32.

36. Feldman S. Varicella-Zoster virus pneumonitis. Chest 1994; 106 (Suppl): 225-275.

37. Haake DA, Zakowski PC, Haake DC, et al. Early treatment with acyclovir for varicella pneumonia in otherwise healthy adults: a retrospective controlled study and review. Rev Infect Dis 1990; 12: 788-798.

38. Riebwasser JH, Harris RE, Bryant RE, et al. Varicella pneumonia in adults: report of 7 cases and a review of the literature. Medicine 1967; 46: 409-423.

39. Esmonde T, Herdman G, Anderson G. Chickenpox pneumonia: an association with pregnancy. Thorax 1989; 44: 812-15.

40. Joseph CA, Noah ND. Epidemiology of chickenpox in England and Wales 1967-1985. Br Med J 1988; 296: 673-676.

41.   Schoub BD. Chickenpox in adults. Virus SA 1992; 1: 1-3.
42.   Mer M, Richards GA. Corticosteroids in life-threatening varicella pneumonia. Chest 198; 114: 426-431.
43.   Fraisse P, Faller M, Rey D, et al. Recurrent varicella pneumonia complicating an endogenous reactivation of chickenpox in an HIV-infected adult patient. Eur Respir J 1998; 11: 776-778.
44.   Stugo I, Israele V, Wittek AE, et al. Clinical manifestations of varicella-zoster virus infections in human immunodeficiency virus-infected children. Am J Dis Child 1991; 147: 742-745.
45.   Popara M, Pendle S, Sacks L, Smego RAJ, Mer M. Varicella pneumonia in patients with HIV/AIDS. Int J Infect Dis 2002; 6: 6-8.
46.   Tibbs C, Williams R. Viral causes and management of acute liver failure. J Hepatitis 1995; 22 (Suppl 1) 68-73.
47.   Bernuau J, Rueff B, Benhamon J-P. Fulminant and sub-fulminant liver failure: definition and causes. Semin Liver Dis 1986; 6: 97-106.
48.   Summerfield JA. Virus hepatitis update. J R Coll Phys London 2000; 34: 381-385.
49.   Skidmore SJ. Factors in spread of hepatitis E. Lancet 1999; 354: 1049-1050.
50.   Acharya S K, Dasarathy S, Kumer T, et al. Fulminant hepatitis in a tropical population: Clinical course, cause and early predictors of outcome. Hepatology 1996; 23: 1448-1455.
51.   Sallie R, Tibbs C, Silva A, et al. Detection of hepatitis E but not C in sera of patients with fulminant hepatitis. Hepatology 1991; 14: 68a.
52.   Emond J, Aran P, Whitington PF, et al. Liver transplant in the management of fulminant hepatic failure. Gastroenterology 1989; 96:1583-1588.
53.   Abramson O, Dagan R, Tal A, Sofer S. Severe complications of measles requiring intensive care in infants and young children. Arch Pediatr Adolesc Med 1995; 149: 1237-40.
54.   Loukides S, Panagou P, Kolokouris D, Kalogeropoulos N. Bacterial pneumonia as a superinfection in young adults. Eur Respir J 1999; 13: 356-60.
55.   Wong RB, Goetz MB. Clinical and laboratory features of measles in hospitalized patients. Am J Med 1993; 95:377-383.
56.   Hussey GD, Klein M. A randomized, controlled trial of Vitamin A in children with severe measles. N Engl J Med 1990; 323: 160-164.
57.   Levitz RE. Herpes simplex encephalitis: A review. Heart Lung 1998; 27: 209-212.
58.   Whitley RJ. Viral encephalitis. N Engl J Med 1990; 323-342.
59.   Klein R, Hirsch MS. Herpes simplex virus Type I encephalitis. Up to Date 2000; 8: 1-9.
60.   Lakeman FD, Whitley RJ. Diagnosis of herpes simplex encephalitis: application of a polymerase chain reaction to cerebrospinal fluid from brain biopsied patients and correlation with disease. National Institute of Allergy and Infectious Diseases. Collaborative Antiviral Study Group. J Infect Dis 1995; 17: 857-863.
61.   Aslanzadeh J, Skiest DJ. Polymerase chain reaction for detection of herpes simplex virus encephalitis. J Clin Pathol 1994; 47: 554-71.
62.   Rubin H. Impact of cytomegalovirus infection on organ transplant recipients. Rev Infect Dis 1990; 12 (Suppl 7): 754- 766.
63.   Stratta RJ, Shaefer MS, Markin RS, et al. Clinical patterns of cytomegalovirus disease after liver transplantation. Arch Surg 1989; 124: 1443-1450.
64.   Ho M. Observations from transplantation contributing to the understanding of pathogenesis of CMV infection. Transplant Proc 1991; 23 (Suppl 3): 104-109.

65. Winston DJ, Ho WG, Champlin RE. Cytomegalovirus infection after allogeneic bone marrow transplantation. Rev Infect Dis 1990; 12 (Suppl 7): 776-797.
66. Chernoff A, Snydman D. Viral infections in the intensive care unit. New Horiz 1993; 1: 279-301.
67. McKenzie R, Travis W, Dolan S, et al. The causes of death in patients with Human Immunodeficiency Virus Infection. A clinical and pathologic study with emphasis on the role of pulmonary diseases. Medicine 1991; 70: 326-343.
68. Wallace J, Hannah J. Cytomegalovirus in patients with AIDS. Chest 1987; 92: 198-203.
69. Miles P, Baughman R, Linneman C. Cytomegalovirus in lavage fluid of patients with AIDS. Chest 1990; 97: 1072-1076.
70. Papazian L, Fraisse A, Garbe L, et al. Cytomegalovirus. An unexpected cause of ventilation associated pneumonia. Anaesthesiology 1996; 84: 280-287.
71. Holladay RC, Campbell GD. Nosocomial viral pneumonia in the intensive care unit. Clin Chest Med 1995; 16: 121-133.
72. van den Berg AP, Klompmaker IJ, Haagsma EB, et al. Antigenemia in the diagnosis and monitoring of active cytomegalovirus infection after liver transplantation. J Infect Dis 1991; 164: 265-270.
73. van den Berg AP, van der Bij W, van Son WJ, et al. Cytomegalovirus antigenemia as a useful marker of symptomatic cytomegalovirus infection after renal transplantation – a report of 130 consecutive patients. Transplantation 1989; 48: 991-995.
74. The TH, van der Ploeg M, van der Berg AP, et al. Direct detection of cytomegalovirus in peripheral blood leukocytes. A review of the antigenemia assay and polymerase chain reaction. Transplantation 1992; 54: 193-198.
75. Hecht DW, Snydman DR, Crumpacker CS, et al. Ganciclovir for the treatment of renal transplant – associated primary cytomegalovirus pneumonia. J Infect Dis 1988; 157: 187-190.
76. Cantarovich M, Hiesse C, Lantz O, et al. Treatment of cytomegalovirus infections in renal transplant recipients with 9- (1,3 – dihydroxy-2-propoxymethyl guanine). Transplantation 1988; 45: 1139-1141.
77. Erice A, Jordan MC, Chace BA, et al. Ganciclovir treatment of cytomegalovirus disease in transplant recipients and other immunocompromised hosts. JAMA 1987; 257: 3082-308.
78. Reed EC, Wolford JL, Kopecky KJ, et al. Ganciclovir for the treatment of cytomegalovirus gastroenteritis in bone marrow transplant patients. Ann Intern Med 1990; 112: 505-510.
79. Shepp DH, Dandliker PS, de Miranda P, et al. Activity of 9-[2-hydroxy-1-(hydroxymethyl) ethoxymethyl] guanine in the treatment of cytomegalovirus pneumonia. Ann Intern Med 1985; 103: 368-373.
80. Reed EC, Bowden RA, Dandliker PS, et al. Treatment of cytomegalovirus with ganciclovir and intravenous cytomegalovirus immunoglobulin in patients with bone marrow transplants. Ann Intern Med 1988; 109: 783-788.
81. Schmidt GM, Kovacs A, Zaia JA, et al. Granciclovir immunoglobulin combination therapy for the treatment of human cytomegalovirus – associated interstitial pneumonia in bone marrow allograft recipients. Transplantation 1988; 46: 905-907.
82. Lee N, Hui D, Wu A et al. A major outbreak of severe acute respiratory syndrome in Hong Kong. N Engl J Med 2003; 348: 1986-1994.
83. Centres for Disease Control and Prevention Update: Outbreak of severe acute respiratory syndrome-world-wide 2003. Morb Mortal Wkl Rep 2003; 52: 241-248.

84. Reilley B, Van Herp M, Sermand D, Dentico N. SARS and Dr Carlo Urbani. N Engl J Med 2003: 348; 1951.

85. Drosten C, Gunther S, Preiser W et al. Identification of a novel coronavirus in patients with severe acute respiratory syndrome. N Engl J Med 2003; 348: 1967-76.

86. Ksiazek T, Endman D, Goldsmith C et al. A novel coronavirus associated with severe acute respiratory syndrome. N Engl J Med 2003; 348: 1953-1966.

87. http://www.who.int/csr/sars/primers/en/

88. http://www.who.int/csr/sars.casedefinitions/en/,

89. Peiris JS, Chu C, Cheng V et al. Clinical progression and viral load in a community outbreak of coronavirus-associated SARS pneumonia: a prospective study. Lancet 2003; 361: 1767-1772.

90. Booth C, Matukas L, Tomlinson et al. Clinical features and short term outcomes of 144 patients with SARS in the greater Toronto area. JAMA 2003; 3289: 2801-2809.

91. So L, Lau A, Yam L et al. Development of a standard treatment protocol for severe acute respiratory syndrome. Lancet 2003; 361:1615-1617.

92. Cheung CY, Poon LL, Lau AS etal. Induction of post inflammatory cytokines in human macrophages by influenza A (H5N1) viruses: a mechanism for the unusual severity of human disease. Lancet 2002; 1831-1837

93. Ning Q, Brown D, Parodo J et al. Ribavirin inhibits viral induced macrophage production of TNF, IL-1, the procoagulant fgl2 prothrombinase and preserves Th1 cytokine production but inhibits Th 2 cytokine response. J Immunol 1998; 160; 3487-93.

94. Wenzel R, Edmond M. Managing SARS amidst uncertainty. N Engl J Med 2003; 348:1947-1948.

95. http://www.who.int.csr/sarsarchive/.

96. Tomlinson B, Cockram C.SARS: experience at Prince of Wales Hospital, Hong Kong. Lancet 2003;361;1486-1487

97. Fisher D, Chew M, Lim Y-T, Tambyah P. Preventing local transmission of SARS: lessions from Singapore. Published on-line ahead of print. Med J Austral 2003. http://www.mja.com.au/public/rop/fis10245.m.html

98. Cooke F, Shapiro D. Global outlook of severe acute respiratory syndrome (SARS). Int J Infect Dis 2003; 7: 80-85.

99. http://www.cdc.gov/ncidod/sars/infection control.htm

100. Seto WH, Tsang D, Yung RW et al and advisors of expert SARS group of Hospital Authority. Effectiveness of precautions against droplets and contact in prevention of nosocomial transmission of severe acute respiratory syndrome (SARS). Lancet 2003; 361: 15191-20.

101. Health Canada Infection Control guidance for respirators (masks) worn by health care workers – frequently asked questions, revised April 17: 2003. http://www.hc.sc.gc.ca/pphb.dgspsp/sars.sras/ic.ci/sars-respmasks.e.html.

102. US Centers for Disease Control: Guidelines for isolation precautions in hospitals. Am J Infect Control 1996; 24: 24-52.

4

Viral Infections in ICU Patients

David E. Greenberg[1] and Stephen B. Greenberg[2, 3]
NIAID[1], Laboratory of Clinical Infectious Diseases, Bethesda, Maryland 20892
Department of Medicine[2] and Department of Molecular Virology and Microbiology[3]
Baylor College of Medicine, Houston, Texas 77030

1 INTRODUCTION

Intensive care units (ICU) were used originally for mechanical ventilation of patients with poliomyelitis and for recovery of patients after anesthesia. Today, these units have been expanded to care for severely ill patients with a wide variety of clinical conditions requiring close monitoring and support. A variety of medical conditions and physiologic disturbances have benefited from this care. Many critically ill patients have underlying infections and of those with infections, viruses are the cause in a small but important percentage (1,2). The majority of viral infections which require care in an intensive care unit involve the respiratory or the central nervous system. However, other organ systems, such as the gastrointestinal tract, may be severely affected by viruses and require support or close monitoring. The conditions reviewed in this chapter are found in adults and do not include HIV infections. Table 1 is a summary of acute illnesses which may be caused by viruses and require treatment in an ICU in the Western Hemisphere. The syndromes and special hosts that are associated with severe viral infections may be diagnosed from epidemiologic clues and specific laboratory tests (3,4). Clinical signs and symptoms are rarely sufficient to make a specific diagnosis of a viral infection.

2 ACUTE RESPIRATORY FAILURE

2.1 Hypoxic → Viral Pneumonia

Although severe community-acquired pneumonia is usually caused by bacteria, viruses account for approximately 3-10% of cases in large series (5,6,7,8). The usual viral causes of pneumonia in adults are influenzavirus type A and B, parainfluenzaviruses, respiratory syncytial virus (RSV), and adenoviruses (9,10,11). These pneumonias may be nosocomially acquired, especially during peak respiratory periods (12,13,14,15,16).

Viruses can cause an atypical pneumonia in otherwise healthy individuals or a pneumonia in immunocompromised hosts (17,18,19,20). The most common cause of viral pneumonia in adults is influenzavirus type A and B. Immunocompromised patients are more likely to have viral pneumonias caused by RSV, cytomegalovirus (CMV), herpes simplex virus (HSV), varicella-zoster virus (VZV), adenovirus and rarely measles (21-35). Radiographic findings are variable and not virus specific. Computed tomographic findings are also variable and overlap by virus etiology. Viral pneumonia may have poorly defined nodules and patchy areas of peribronchial ground-glass opacity and air-space consolidation. Hyperinflation is common (36).

Influenzavirus pneumonia is more likely to occur in the elderly and immunocompromised patient. Predisposing conditions for influenza pneumonia include age >65 years, diabetes mellitus, chronic lung disease, pregnancy, and immunosuppression. Influenzavirus pneumonia can be overwhelming and rapidly fatal (37-39). Influenzavirus infection can lead to bacterial superinfection and pneumonia secondary to *Streptococcus pneumoniae* and less commonly, *Staphylococcus aureus*. Diagnosis can be made by obtaining respiratory secretions and testing for viral antigen or virus growth in cell culture, or by serologic tests such as HI or ELISA.

Viral Infections in the ICU

Table 1. Virus Infections in ICU Patients

Syndrome	Viruses Associated
•Acute Respiratory Failure	
Hypoxic → Pneumonia	Influenza A & B, RSV, coronavirus (SARS), adenovirus, CMV, VZV, HSV
Hypercapnic – Hypoxic →	
COPD/Asthma exacerbation	Influenza A & B, coronavirus, rhinovirus, adenovirus
Acute Respiratory Distress Syndrome	Influenza A & B, coronavirus, rhinovirus
Without lung disease	
Guillain-Barré Syndrome	HSV, CMV, EBV
	Hantavirus, influenzavirus
•Shock	
Cardiogenic → myocarditis	Enteroviruses
Distributive → "sepsis"	Dengue Fever, VHF
•Altered Mental Status	
Meningitis/Encephalitis	HSV, enteroviruses, arboviruses, West Nile Virus, rabies
Fulminant hepatitis	Hepatitis B, D, E
Rhabdomyolysis	Influenza A & B, CMV
Acute Pancreatitis	Mumps, parainfluenza, enterovirus
•Bioterrorism Agents	Smallpox, VHF, encephalitides
Special or Immunocompromised Host	
•Burns/Trauma	HSV, CMV
•Pregnancy	VZV, CMV, HSV, influenzavirus
•Transplantation	CMV, HSV, RSV, influenzavirus A and B, HHV-6, BKV, adenovirus

Although there are several antiviral agents which are effective against influenzavirus, there are few controlled studies on treatment of influenzavirus pneumonia. The older drugs, amantadine and rimantadine are useful against influenzavirus type A only and are only available in oral formulation. The newer neuraminidase inhibitors, oseltamivir and zanamivir are effective against influenzavirus types A and B, and are also only available in oral form (40,41). Ribavirin can be given intravenously and has been shown to be effective in a few published cases of influenzavirus as well as RSV pneumonia (42-44).

Recent outbreaks of severe acute respiratory syndrome (SARS) have been published which have been associated with a new coronavirus   (45). A case fatality rate of 4% to 15% has been reported in series from China and Canada (46). SARS should be considered in patients with the following history: travel in the previous 10 days to places with documented or a suspected SARS case; or close contact with a person with known or suspected SARS. Clinical illness includes cough, shortness of breath, hypoxia, and radiographic evidence of pneumonia or ARDS without another cause (47-50). Because SARS has been shown to transmit to healthcare personnel, isolation of patients is required. Standard, contact, and airborne isolation precautions should be initiated in suspected cases. All cases require notification to the local public health department and the Centers for Disease Control.

2.2 Hypercapnic-Hypoxic

2.2.1 COPD/Asthma Exacerbation. Acute respiratory failure can occur in patients with chronic obstructive pulmonary disease (COPD) and lead to hospitalization and the need for mechanical ventilation (51). Patients with exacerbations of COPD often have a history of increased sputum production, dyspnea and cough. Documented viral infections occurs in up to 45% of episodes of exacerbation of COPD (52,53). The most frequently identified viruses in acutely ill COPD patients are rhinoviruses, parainfluenzaviruses, coronaviruses, and influenzaviruses type A and B (53). In severely ill adult patients requiring hospitalization and mechanical ventilation, influenzaviruses and coronaviruses are most common. The epidemiology of acute exacerbations in adult asthmatic patients is similar. With acute, severe, exacerbations of asthma, patients may require close monitoring and ventilatory support. Recent studies have confirmed the importance of rhinoviruses, parainfluenzaviruses, and influenzaviruses in over half the acute exacerbations in adult asthmatics (54).

2.2.2 Acute Respiratory Distress Syndrome (ARDS). ARDS is characterized by severe hypoxemia and diffuse infiltrates on chest x-ray in the absence of clinical heart failure. ARDS has many causes. Common illnesses associated with ARDS include sepsis, diffuse pneumonia, aspiration or severe trauma. Although bacteria are more often the cause of infections leading to ARDS, viruses have been identified, especially influenzavirus, Hantavirus, varicella, herpes simplex virus and SARS (55-57).

Hantaviruses cause very different illnesses in Europe and Asia as opposed to the Americas (58,59). Epidemics of hemorrhagic fever with renal syndrome (HFRS) are caused by Hantaan, Dobravna, and Seoul viruses. Hantavirus pulmonary syndrome (HPS) is spread by New World rats and mice carrying one of the dozen reported viruses of the Bunyavirus family. Sin Nombre virus is the major pathogen causing HPS in the United States (60). HPS causes a non-cardiogenic pulmonary edema after initial influenza-like symptoms. HPS has a higher mortality than HFRS. The virus is transmitted to humans by inhalation of aerosolized particles of rodent excreta or by direct rodent contact. HPS has been recognized in 31 states, most commonly in the western region of the United States. Laboratory abnormalities include thrombocytopenia, leukocytosis, hemoconcentration, and the presence of immunoblasts on peripheral smear (61,62). Diagnosis requires serologic detection of IgM and IgG antibodies to Sin Nombre virus. Mortality rates may approach 40%. Although there is no specific therapy, cardiopulmonary support in an ICU may be associated with improved survival.

2.3 Without Lung Disease.

Guillain-Barré Syndrome (GBS) is the most common cause of acute neurogenic respiratory failure. Ventilatory assistance will be necessary in up to one-third of patients with this acute inflammatory disseminated polyneuropathy (AIDP). If respiratory failure occurs, it usually does so during the first two weeks of the illness. An antecedent respiratory or gastrointestinal illness is commonly reported in the preceding four weeks prior to the onset of symptoms (63,64,65). Viruses commonly reported in GBS patients include influenzavirus, CMV, acute and chronic hepatitis B, EBV, and varicella-zoster virus (66-75). Rare cases have been reported and associated with West Nile virus, Parvovirus B19, Hantavirus, rubella, and dengue (76-80).

The care of these patients in ICUs has significantly reduced mortality rates. In patients requiring mechanical ventilation, plasma exchange has shortened the duration of respiratory support (81). An alternative treatment to plasma exchange appears to be IVIg. Even with treatment, the rehabilitation process is long and residual weakness is found in approximately 15%.

3 SHOCK

3.1 Cardiogenic → Myocarditis

Cardiogenic shock may develop following acute myocardial infarction or severe heart failure from any cause. Clinically, cardiogenic shock is manifested by peripheral hypoperfusion, cold extremities, cyanosis, or hypotension. Viral infection of the myocardium can lead to clinical myocarditis which is severe enough to manifest cardiogenic shock either due to myocardial failure or tachyarrhythmias (82,83).

Although rarely proven, viruses are suspected as the major causes of acute myocarditis. A suspicion of viral myocarditis should be high in patients who had fever and myalgias preceding the development of cardiac symptoms. Evidence of myocardial damage with elevated creatine kinase and troponin levels is common. Patients may present with typical anginal chest pain and/or arrhythmias; making it difficult to rule out an acute myocardial infarction (84).

The viruses documented in myocarditis cases are predominantly enteroviruses, especially coxsackie B viruses (85). However, influenzavirus, adenoviruses, Parvovirus B19, and CMV have been reported, as well (86-92). Recent smallpox vaccine studies have also reminded us of the potential for myocarditis with vaccinia virus. Most patients with acute myocarditis have mild heart failure, but occasionally, the myocarditis and associated organ dysfunction is severe enough to require mechanical ventricular assist devices until resolution or cardiac transplantation is available.

3.2 Distributive "Sepsis"

3.2.1 Dengue Fever. Dengue hemorrhagic fever (DHF), and dengue shock syndrome (DSS) have become increasingly prevalent in the past decade (93-95). The principal vector, *Aedes aegypti*, as well as *Aedes albopictus*, or the tiger mosquito, was brought to Houston, Texas from Japan in the 1980's. Clinically, dengue fever presents with high fever, severe headaches, myalgias, joint pains, maculopapular rash and leukopenia. Petechiae secondary to thrombocytopenia can be observed.

DHF/DSS are severe forms of the infection. The main pathophysiologic changes seen in DHF/DSS patients are abnormal hemostasis and plasma leakage.

With DHF/DSS, patients present with high fever, facial flushing and headache. Hemorrhagic manifestations such as bleeding at venipuncture sites, petechiae, epistaxis, gingival bleeding, and gastrointestinal hemorrhage may be observed. The tourniquet test is often positive. After the patient becomes afebrile, profuse sweating and a drop in blood pressure are observed. Once shock develops, the patient's condition deteriorates and death may occur without support. Recent series have reported DSS in the absence of thrombocytopenia and hemoconcentration. Mortality rates vary from 2-10% for DHF.

Risk factors for DHF/DSS in the Americas include 1) secondary infection with a different serotype; 2) sequential serotypes in secondary infections; 3) association with dengue 2 virus and, less frequently, dengue 3 virus; 4) longer interval between first and second infection; 5) young age; 6) lower frequency in blacks; and 7) individuals with a history of asthma, diabetes mellitus, or sickle cell anemia (93). To consider a case, the travel history and local disease occurrence must be known. Diagnosis is made by detection of IgM antibody. Paired sera for the testing of specific rises in IgG antibody will give a more specific diagnosis. Treatment for DHF/DSS is supportive.

3.2.2 Viral Hemorrhagic Fevers (VHF). Viral hemorrhagic fevers (VHF) are caused by RNA viruses and transmitted by rodent or arthropod vectors. In the case of Marburg and Ebola viruses, the reservoir and mode of transmission remain unknown. Case-fatality rates vary from 1% to 90%. Clinical syndromes include hemorrhage secondary to capillary leakage, hepatitis, encephalitis, and/or nephropathy. Disseminated intravascular coagulopathy (DIC) is common to many but not all of these viruses. The viruses found naturally in the Western hemisphere include yellow fever, dengue, Sin Nombre virus (Hantavirus), and South American hemorrhagic fever viruses (Guanarito, Sabia, Junin, and Machupo) (Table 2). Other viruses indigenous to Africa and Asia include Rift Valley fever, Crimean-congo hemorrhagic fever, Kyasanur Forest virus, Omsk hemorrhagic fever, hemorrhagic fever with renal syndrome, Lassa fever, and African hemorrhagic fever (Marburg and Ebola). The hemorrhagic fever viruses in Asia and Africa have the potential for introduction into the Western hemisphere or for use as biological warfare agents.

Table 2. Viral Hemorrhagic Fevers (VHF) in the Western Hemisphere

	Flaviviridae	Bunyaviridae	Arenaviridae
Disease	Yellow fever Dengue HF	Hantavirus Pulmonary Syndrome	Venezuelan, Brazilian, Argentine, & Bolivian HF
Clinical Syndromes	Hemorrhage, hepatitis, nephropathy	ARDS, hepatitis, nephropathy	Hemorrhage, encephalitis, nephropathy
Case-fatality Rate	2-20%	40-50%	10-33%
Treatment	Supportive care, close monitoring, no aspirin, no steroids, fluid resuscitation	Oxygenation and monitoring crystalloid fluid replacement, ribavirin IV (experimental)	Supportive therapy, specific immune human plasma (Argentine HF)

The viruses causing hemorrhagic fever cause microvascular damage and alterations in vascular permeability (96,97). Fever, prostration, and myalgias are common initial symptoms. Physical examination will often reveal hypotension, facial flushing, and petechiae. With progression to shock and generalized bleeding, there is often hematopoietic, neurologic and pulmonary involvement. Hepatitis is common and severe with yellow fever but not with the other VHF agents of the Western hemisphere. Increasing vascular permeability, loss of intravascular volume, and multiorgan failure are often the final pathway to death.

Routine laboratory tests yield nonspecific abnormalities, but thrombocytopenia and coagulation defects should suggest the diagnosis with the corresponding clinical presentation. Specific diagnosis requires isolation of the virus or staining of formalin-fixed tissue. Only the CDC and USAMRIID laboratories are under Biolevel safety 4 conditions which are necessary for attempting isolation of these viruses. RT-PCR assays for rapid diagnosis may become more widely available in the future.

The care of patients with VHF is largely supportive. Patients in shock and actively bleeding will require close monitoring, fluid resuscitation as well as transfusion of red cells, platelets, and clotting factors (98,99). No aspirin or corticosteroids should be used. Dopamine may be the pressor of choice in unresponsive shock. Only ribavirin has been used successfully in VHF, especially for Lassa fever patients. Experimental use of ribavirin in Hantavirus pulmonary syndrome is being evaluated. Specific immune human plasma has been successful in treating Argentine hemorrhagic fever and may be useful in Bolivian hemorrhagic fever.

All suspected VHF cases should be placed in isolation immediately. Respiratory precautions and placement in a negative pressure room should help reduce the spread to hospital personnel and close contacts. All specimens must be appropriately labeled. Decontamination of all areas where the patient has been is essential. Reporting all suspected cases to local, state, and federal authorities is necessary to alert the community of a possible outbreak.

4 ALTERED MENTAL STATUS

4.1 Viral Meningitis and Encephalitis
The differential diagnosis of viral meningitis and encephalitis includes a long list of viruses as well as bacteria and noninfectious conditions. Because the clinical presentation of viral meningitis and encephalitis overlaps with other infections and illnesses, the diagnostic evaluation and therapeutic management are complicated. The most common viral pathogens associated with meningitis or encephalitis and their epidemiology are listed in Table 3 (100-103).

Viral meningitis and encephalitis cases occur worldwide. In the United States, echoviruses and coxsackie viruses (non-polio enteroviruses) are the cause of many cases of meningitis and some cases of encephalitis. (104). Although outbreaks are more common in the summer months, CNS enteroviral illness occurs throughout the year. The most common cause of sporadic cases of encephalitis in adults is HSV-1. Over 50% of the cases of encephalitis in adults older than 50 years are caused by HSV-1 (105). Outbreaks of encephalitis can be caused by alphaviruses and/or flaviviruses such as St. Louis Encephalitis. West Nile virus infection and illness has only been reported in the United States since 1999 (106). However, severe CNS infections and deaths from West Nile virus are being reported with increased frequency throughout many states during 2002 and 2003 (107). Rare cases of rabies are reported annually in the United States, but the mortality rate in these patients approaches 100% (108,109).

The clinical presentation of viral meningitis and encephalitis often includes fever, headache, vomiting, altered mental status, and seizures. On neurologic examination, hyperreflexia, cognitive alterations, ataxia, and focal findings are common. Examination of the cerebrospinal fluid is necessary for evaluating these patients and sending specimens for diagnosis. Typically, patients with viral meningitis and encephalitis have elevated protein levels, normal glucose levels and a lymphocytosis. In some cases, there may be a neutrophil pleocytosis and/or mild hypoglycorrhachia. These findings will often prompt clinicians to begin broad-spectrum antibiotics as well as antivirals until culture and PCR results are known. Magnetic resonance imaging (MRI) can demonstrate a pattern that suggests a specific pathogen such as HSV-1 or Japanese encephalitis. In other cases, the MRI may be normal. In herpes simplex encephalitis (HSE), 95% of patients have a history of change in the level of consciousness, 90% have fever, 85% have personality change, and 80% have headaches. Seizures have been reported in approximately two-thirds of patients. Other signs and symptoms include aphasia, amnesia and/or hallucinations. Typically, the CSF profile includes elevated opening pressure, lymphocytes, red blood cells and slightly elevated protein level. After 18 to 24 hours of symptoms, approximately 95% of CSF samples will have a positive PCR test (110). Serum antibody assays are not helpful in diagnosis. EEG shows lateralized epileptiform discharges (PLEDs) over the temporal or frontal area in 85% of patients with HSE. In the first 5 to 6 days of illness, the CT scan maybe normal. MRI is more likely to be abnormal earlier than the CT scan. Brain biopsy, although the definitive diagnostic procedure, is only performed when there is an atypical clinical, radiologic, or laboratory picture. The mortality rate has been reduced to 20% with acyclovir therapy. However, the mortality rate is higher in those older than 30 years and in those who were comatose before treatment was initiated.

West Nile virus had been a cause of encephalitis in Africa and the Middle East until 1999, when it was found in fatal cases of encephalitis in New York City (111,112). Over the next few years, hundreds of cases had been confirmed throughout most of the United States. West Nile virus is a member of the Flaviviridae family. Most infected patients are asymptomatic. Clinical illness is associated with fever, headache, anorexia and malaise. Lymphadenopathy has been reported in earlier epidemics. A maculopapular rash over the chest, back, and upper extremities is observed in approximately 50% of cases. Other signs and symptoms include conjunctivitis, nausea, abdominal pain, and diarrhea (113,114).

Meningitis, encephalitis, and acute flaccid paralysis have been reported with West Nile virus (115-117). CSF examination usually reveals a lymphocytic pleocytosis and elevated protein level. Guillain-Barré syndrome has also been reported. Diagnosis can be made by serologic tests, but cross-reactive antibodies to other flaviviruses have been observed. Detection of virus by RT-PCR is positive in less than 50% of cases.

Patients may need prolonged ventilation because of neurologic complications. No specific therapy has been found to be effective, although interferon, ribavirin and IVIg have been tested or proposed for clinical trials. Recovery may be prolonged with neurologic abnormalities being reported for up to one year following the illness.

Specimens should be collected for viral isolation and PCR testing. CSF PCR tests are available for detecting HSV-1, CMV, EBV, VZV, HHV-6, HHV-7, and the enteroviruses. In HSV-1 encephalitis, the sensitivity of CSF PCR approaches 95% (118). Less information is available on the sensitivity of CSF PCR tests for the other viruses.

For many patients with meningitis, admission to the hospital will not be necessary. Patients with suspected encephalitis should be admitted to the hospital and monitored closely. Initiation of empiric acyclovir intravenously should be considered. An MRI should then be reviewed for evidence of mass effect. If significant mass effect is present, then the CSF examination should be deferred and evidence of increased intracranial pressure (ICP) would require specific treatment (119). If no mass effect is found, then CSF examination should be performed and the appropriate tests ordered. If the HSV PCR test is negative, consider other diagnoses unless MRI is compatible with HSV.

Table 3. Epidemiology of Viral Encephalitis/Meningitis in the Western Hemisphere

Virus	Clinical Findings	Epidemiology
Enteroviruses	Myocarditis, rash pleurodynia	Summer and fall peaks both epidemic; sporadic cases
Herpes Simplex Type 1	Focal neurologic deficits	No seasonality; sporadic
Herpes Simplex Type 2	Primary genital lesions	No seasonality; sporadic
Varicella-Zoster virus	Rash	No seasonality; sporadic
EBV/CMV	Mononucleosis syndrome, immunosuppressed	No seasonality; sporadic
HHV-6	Focal neurologic	No seasonality; sporadic
EEE/WEE/VEE	Mosquito-borne, seizures	Summer; epidemic
SLE/WNV	Mosquito-borne, SIADH, acute flaccid paralysis	No seasonality; sporadic
California	Seizures	Northern Central US; summer; epidemic
Influenzavirus	Pneumonia	Winter; epidemic
LCM	---	Contact with rodents; winter; sporadic
Mumps	Parotitis	Spring/summer; sporadic
Rabies	Hydrophobia	History of animal bite; no seasonality; sporadic

5 FULMINANT HEPATITIS

In 1% to 2% of patients with acute hepatitis, acute liver failure or fulminant hepatitis will occur (120). Patients with fulminant hepatitis will often manifest encephalopathic changes, jaundice, and a prolonged prothrombin time. Levels of elevated aminotransferases do not have prognostic value and may decrease as the liver failure becomes worse. The viruses reported in cases of fulminant hepatitis are hepatitis B and D, A and E and rarely C (121,122). A recent study reports that approximately 12% of acute liver failure causes are attributable to viral hepatitis (123). Occasional cases of herpes simplex and varicella hepatitis have been reported in immunocompetent and immunosuppressed adults (124-126). A recent study has found HHV-6 antigens in explanted livers of patients with acute liver failure of unknown etiology (127).

6 RHABDOMYOLYSIS

Rhabdomyolysis is a syndrome characterized by muscle weakness, pain and cramps. Complications may be extramuscular and lead to acute renal failure and severe metabolic abnormalities (128,129). Rhabdomyolysis may be purely related to exertion, genetically determined or due to nonhereditary, nonexertional causes. Viruses have been reported to be rare precipitative causes of rhabdomyolysis (130,131). These viruses have been cytomegalovirus, measles, adenovirus, varicella, EBV, influenzavirus, parainfluenzavirus and enteroviruses (132-150).

7 PANCREATITIS

Acute pancreatitis is characterized by abdominal pain and elevated serum amylase and lipase (151). Depending upon the severity of the pancreatitis, the patient may need mechanical ventilation and support in an intensive care unit. The majority of cases are caused by gallstones and alcohol. Pancreatitic obstruction, drugs and toxins, metabolic and genetic disorders, trauma and iatrogenic causes account for most of the remaining cases. Several viruses are rare causes of acute pancreatitis: mumps, coxsackievirus, cytomegalovirus, hepatitis B, varicella-zoster virus, herpes simplex virus, hepatitis C, enterovirus 71, hepatitis E, hepatitis A, and adenovirus (152-160).

8 BIOTERRORISM AGENTS

The possible use of biological agents in a terrorist attack has become quite real since September 11, 2001 and the subsequent distribution of anthrax (161,162). Potential biological weapons have included both bacteria and viruses. The chief viral candidates with the greatest impact are smallpox, viral hemorrhagic fever agents such as Ebola virus, and equine encephalitis (163,164). Delivery of the virus weapon is likely to be by the respiratory route; contamination of food or water is less likely.

8.1 Smallpox

A single case of smallpox would have significant impact on the health care system (165). Smallpox is caused by variola virus and is highly infectious. Spread may occur by person-to-person or by fomites. After a 12 to 14 day incubation period, a prodromal illness marked by fever, rigors, malaise, headache and backache will last approximately three days. During this time, the physician examining the patient will consider a "flu-like" illness is most likely (166). A discrete centrifugal rash characterized by macules begins on the face, hands, and forearms (167). The macules become papules and then vesicles which spread over the whole body. Pustules and crusted lesions develop by the eighth day and thirteenth day, respectively. The rash can be distinguished from varicella (chickenpox) by being more peripheral and progressing at the same stage over the skin.

Mortality may reach 30% and is highest during the second week of illness. Pulmonary edema and hemoptysis were commonly reported in earlier outbreaks. Renal failure and electrolyte abnormalities also contributed to the morbidity in smallpox patients.

If a suspected case of smallpox were admitted to the hospital or ICU, the physician would need to notify local and national public health officials. Specimens should be sent for diagnosis to state and national laboratories where biosafety level 4 precautions are available.

A suspected case of smallpox needs to be placed in strict airborne and contact isolation and in a negative-pressure room. Patient transport should be limited. Dedicated equipment should be used. Linens from the patient should be autoclaved before laundering (Table 4). Isolation of the patient should be continued until all scabs separate.

Smallpox patients will need monitoring and excellent skin care. There is no currently approved antiviral treatment, although cidofovir has been shown to be effective in vitro and in animal models (168,169). Vaccination of contacts should limit the development of clinical disease if administered within four days of exposure (170).

8.2 Viral Hemorrhagic Fevers (VHF)

VHF are caused by a diverse group of viruses that are transmitted by animal and arthropod vectors (171). VHF agents are potential biologic weapons because they are highly infectious, stable as aerosols, and cause high morbidity and mortality. These viral agents can infect by direct contact with needles, fluids and tissues of other infected patients, however aerosol infection in humans has not been reported with the exception of Hantavirus.

All suspected cases of VHF should be placed in isolation in negative-pressure rooms. Airborne, droplet and contact precautions should be instituted. Patient transport should be limited and masks should be placed if transport is essential. Dedicated equipment should be left in the patient's room. Disinfection of surfaces with 10% bleach solution or phenolic disinfectant is recommended. The patient should remain in isolation for the duration of the illness (Table 4).

*Table 4. Isolation Guidelines for Presumed or Documented Viral Infections Related to Bioterrorism**

	Smallpox	VEE*	Viral Enceph.	VHF
Isolation Precautions				
Standard Precautions	X	X	X	X
Contact Precautions (gowns and gloves)	X			X
Airborne Precautions (negative pressure room & N95 masks for all individuals entering rm)	X			
Droplet Precautions (surgical mask)			X	
Patient Placement				
No restrictions				
Cohort "like" patients when private room unavailable	X		X	
Private Room	X	X		X
Negative Pressure	X			
Door closed at all times	X			
Patient Transport				
No restrictions			X	
Limit movement to essential medical purposes only	X	X		X
Place mask on patient to minimize dispersal of droplets		X		
Cleaning, Disinfection of Equipment				
Routine terminal cleaning of room with hospital approv. Disinfectant upon discharge	X	X	X	
Disinfect surfaces with bleach/water sol. 1:9 (10% sol.)				X
Dedicated equipment (disinfect prior to leaving room)	X			X
Linen management as with all other patients	X	X	X	X
Post-mortem Care				
Follow principles of Standard Precautions	X	X	X	X
Droplet Precautions				
Airborne Precautions (negative pressure room & N95 masks for all individuals entering rm)	X			

*Adapted from Karwa M, et al. Bioterrorism and Critical Care. Crit Care Clin 2003; 19:279-313, with permission)
*VEE = Venezuelan Equine Encephalitis

9 SPECIAL OR IMMUNOSUPPRESSED HOSTS

9.1 Burns/Trauma

Burn patients have been reported to have a high incidence of herpes simplex virus (HSV) infections manifested as ARDS and occasionally as pneumonia (172-174). Necrotizing tracheobronchitis, and facial rashes have also been found in these patients (175-179). With acyclovir therapy, there has been reported clinical improvement. Isolation of HSV from bronchoalveolar lavage specimens is associated with the need for assisted ventilation in burn patients (180,181). However, there are conflicting studies on whether HSV activation is associated with increased mortality in these patients (182). A recent prospective study culturing for HSV in the respiratory tract of patients in critical care units showed HSV reactivation was frequent and associated with ARDS and increased length of stay in intensive care (183). This confirmed results from other older studies (184,185). Similar findings have also been found in critically ill surgical patients (186-188).

9.2 Pregnancy

There are numerous physiologic and immunologic changes in pregnancy that create a state of relative immunosuppression (189). Pregnant women are more susceptible to a variety of viral infections including pneumonia (189,190). Influenzavirus, VZV, and measles have all been reported as causes of pneumonia (191-194). Complications of VZV can be particularly devastating during pregnancy. Ninety percent of adults living in nontropical areas are immune to varicella, therefore most pregnant women are not susceptible (195). Pneumonia can occur a few days after the development of rash and fever and can lead to respiratory failure. Both the mother and fetus can suffer morbidity and mortality from varicella. Patients with the complication of varicella pneumonia should be treated aggressively with antiviral therapy and close monitoring (194-197). Acyclovir is generally accepted as the treatment of choice despite the lack of safety studies (198). In pregnant women with no history of VZV who have an exposure, the use of VZIG may be beneficial in preventing maternal infection (199). Another virus that may lead to admission to the ICU is HSV. HSV infection can affect both the mother and the developing fetus (200-202). HSV of the genital tract can be transferred to the newborn and result in severe and life-threatening disease (203). Although rare, HSV type 2 causing encephalitis following cesarean section has been reported (204). Finally, given recent concerns regarding bioterrorism, it is important to realize that smallpox has had high mortality in the pregnant woman compared to the non-pregnant woman (205).

9.3 Transplantation

Viral infections can cause severe morbidity and mortality in both solid-organ and stem cell transplant patients (206-211). These infections can manifest as either primary infection or reactivation of latent disease (212,213). Viral infections in transplant patients may present as a variety of clinical syndromes that require ICU stays. These include pneumonia, encephalitis, hepatitis and gastroenteritis (180,214-219).

The variety of clinical presentations and radiographic features make diagnosis of specific viruses difficult. In the transplant patient, common community respiratory viruses such as influenzaviruses A and B, respiratory syncytial virus (RSV), adenovirus and parainfluenzavirus (PIV) can lead to fulminant pneumonia, respiratory failure and death (4,220-224). There has been increased mortality reported in hematopoietic stem cell transplant patients who develop RSV or PIV infection (225-227). Occasional cases with metapneumovirus have been reported in stem cell transplant recipients (228,229). Both ribavirin with and without IVIG have been used to treat RSV pneumonia; but mortality remains high (33,230,231).Treatment benefit is generally seen when used prior to the development of respiratory failure (232). Other respiratory viruses that can lead to severe pneumonia are rhinoviruses and adenoviruses (233). Solid organ transplant recipients also are at risk of developing severe lower respiratory tract viral infections. One study found a high incidence of influenza among lung transplant patients, but liver and kidney transplant patients also develop influenza (234).

The Herpesviruses cause a variety of clinical syndromes in transplant patients. CMV, EBV, VZV, HSV, HHV-6 and HHV-8 have all been reported to cause disease (188,235-238). CMV disease constitutes a serious problem in bone marrow transplant recipients with a 30-50% incidence of clinically significant infections (239,240). Pneumonitis is the most serious complication and was associated with high mortality prior to antiviral therapy (241). Ganciclovir and foscarnet have been frequently used for prophylaxis and treatment of CMV (242-245). Hyperimmune globulin in combination with antivirals has also been used for treatment of disease (246). Ganciclovir resistance in CMV has been reported and should be considered when clinical responses do not occur (247). EBV can reactivate in transplant recipients and lead to uncontrolled B-cell proliferation and post-transplant lymphoproliferative disorder (PTLD) (248). This can present as frank lymphoma with high mortality rates. Donor lymphocyte infusions and anti-CD20 antibody has been used for treatment (249). HSV can present as both severe mucocutaneous disease or in rare cases encephalitis (250). HSV disease results from reactivation of latent virus (251). Other viruses that cause encephalitis in transplant patients are CMV, VZV, EBV, HHV-6 and recently West Nile Virus (252). Cases of encephalitis usually require treatment in the ICU. In this patient population, viral pneumonia and encephalitis require care in an

ICU setting. Although viral culture has been the gold standard for diagnosis, PCR has become the new diagnostic standard (225,253).

10 REFERENCES

1. Chernoff AE, Snydman DR. Viral infections in the intensive care unit. New Horiz 1993;1:279-301.
2. Couch RB, Englund JA, Whimbey E. Respiratory viral infections in immunocompetent and immunocompromised persons. Am J Med 1997;102(3A):2-9.
3. Greenberg SB. Infections in the immunocompromised rheumatologic patient. Crit Care Clin 2002;18:931-56.
4. Hicks KL, Chemaly RF, Kontoyiannis DP. Common community respiratory viruses in patients with cancer: more than just "common colds". Cancer. 2003;97:2576-87.
5. Dowell SF, Anderson LJ, Gary HE Jr, Erdman DD, Plouffe JF, File TM Jr, Marston BJ, Breiman RF. Respiratory syncytial virus is an important cause of community-acquired lower respiratory infection among hospitalized adults. J Infect Dis 1996;174:456-62.
6. Dudding BA, Wagner SC, Zeller JA, Gmelich JT, French GR, Top FH Jr. Fatal pneumonia associated with adenovirus type 7 in three military trainees. N Engl J Med 1972;286:1289-92.
7. Greenberg SB. Viral pneumonia. Infect Dis Clin North Am 1991;5:603-21.
8. Rello J, Diaz E. Pneumonia in the intensive care unit. Crit Care Med 2003;31:2544-51.
9. Baine WB, Luby JP, Martin SM. Severe illness with influenza B. Am J Med 1980;68:181-9.
10. Luksza AR, Jones DK. Influenza B virus infection complicated by pneumonia, acute renal failure and disseminated intravascular coagulation. J Infect 1984;9:174-6.
11. Motallebi M, Mukunda BN, Ravakhah K. Adenoviral bronchopneumonia in an immunocompetent adult: computed tomography and pathologic correlations. Am J Med Sci 2003;325:285-7.
12. Aitken C, Jeffries DJ. Nosocomial spread of viral disease. Clin Microbiol Rev 2001;14:528-46.
13. Holladay RC, Campbell GD Jr. Nosocomial viral pneumonia in the intensive care unit. Clin Chest Med 1995;16:121-33.
14. Ibrahim EH, Ward S, Sherman G, Kollef MH. A comparative analysis of patients with early-onset vs late-onset nosocomial pneumonia in the ICU setting. Chest 2000;117:1434-42.
15. Raad I, Abbas J, Whimbey E. Infection control of nosocomial respiratory viral disease in the immunocompromised host. Am J Med 1997;102:48-52
16. Valenti WM. "Selected Viruses of Nosocomial Importance." In *Hosptial Infections 4th ed.*, JV Bennett and PS Brachman, ed. Philadelphia PA: Lippincott-Raven, 1998.
17. Whimbey E, Bodey GP. Viral pneumonia in the immunocompromised adult with neoplastic disease: the role of common community respiratory viruses. Semin Respir Infect. 1992;7:122-31.
18. Whimbey E, Champlin RE, Couch RB, Englund JA, Goodrich JM, Raad I, Przepiorka D, Lewis VA, Mirza N, Yousuf H, Tarrand JJ, Bodey GP.

Community respiratory virus infections among hospitalized adult bone marrow transplant recipients. Clin Infect Dis 1996;22:778-82.

19. Whimbey E, Couch RB, Englund JA, Andreeff M, Goodrich JM, Raad II, Lewis V, Mirza N, Luna MA, Baxter B, et al. Respiratory syncytial virus pneumonia in hospitalized adult patients with leukemia. Clin Infect Dis 1995;21:376-9.

20. Whimbey E, Ghosh S. Respiratory syncytial virus infections in immunocompromised adults. Curr Clin Top Infect Dis 2000;20:232-55.

21. Arata K, Sakata R, Iguro Y, Toda R, Watanabe S, Eitsuru Y. Herpes simplex viral pneumonia after coronary artery bypass grafting. Jpn J Thorac Cardiovasc Surg 2003;51:158-9.

22. de la Hoz RE, Stephens G, Sherlock C. Diagnosis and treatment approaches of CMV infections in adult patients. J Clin Virol 2002;25 Suppl 2:S1-12.

23. Eisenstein LE, Cunha BA. Herpes simplex virus pneumonia presenting as failure to wean from a ventilator. Heart Lung 2003;32:65-6.

24. Feldman S, Stokes DC. Varicella zoster and herpes simplex virus pneumonias. Semin Respir Infect 1987;2:84-94.

25. Graham BS, Snell JD Jr. Herpes simplex virus infection of the adult lower respiratory tract. Medicine (Baltimore) 1983;62:384-93.

26. Greenberg SB. Respiratory herpesvirus infections. An overview. Chest 1994;106:1S-2S.

27. Guidry GG, Black-Payne CA, Payne DK, Jamison RM, George RB, Bocchini JA Jr. Respiratory syncytial virus infection among intubated adults in a university medical intensive care unit. Chest 1991;100:1377-84.

28. Klainer AS, Oud L, Randazzo J, Freiheiter J, Bisaccia E, Gerhard H. Herpes simplex virus involvement of the lower respiratory tract following surgery. Chest 1994;106:8S-14S.

29. Manna A, Cordani S, Canessa P, Pronzato P. CMV infection and pneumonia in hematological malignancies. J Infect Chemother 2003;9:265-7.

30. Mohsen AH, McKendrick M. Varicella pneumonia in adults. Eur Respir J 2003;21:886-91.

31. Ramsey PG, Fife KH, Hackman RC, Meyers JD, Corey L. Herpes simplex virus pneumonia: clinical, virologic, and pathologic features in 20 patients. Ann Intern Med 1982;97:813-20.

32. Retalis P, Strange C, Harley R. The spectrum of adult adenovirus pneumonia. Chest 1996;109:1656-7.

33. Stephan F, Meharzi D, Ricci S, Fajac A, Clergue F, Bernaudin JF. Evaluation by polymerase chain reaction of cytomegalovirus reactivation in intensive care patients under mechanical ventilation. Intensive Care Med 1996;22:1244-9.

34. Umans U, Golding RP, Duraku S, Manoliu RA. Herpes simplex virus 1 pneumonia: conventional chest radiograph pattern. Eur Radiol 2001;11:990-4.

35. Zahradnik JM. Adenovirus pneumonia. Semin Respir Infect 1987;2:104-11.

36. Kim EA, Lee KS, Primack SL, Yoon HK, Byun HS, Kim TS, Suh GY, Kwon OJ, Han J. Viral pneumonias in adults: radiologic and pathologic findings. Radiographics 2002;22 Spec No:S137-49.

37. Fleming DM, Zambon M. Update on influenza and other viral pneumonias. Curr Opin Infect Dis 2001;14:199-204.

38. Oikonomou A, Muller NL, Nantel S. Radiographic and high-resolution CT findings of influenza virus pneumonia in patients with hematologic malignancies. AJR Am J Roentgenol 2003;181:507-11.

39. Oliveira EC, Marik PE, Colice G. Influenza pneumonia: a descriptive study. Chest 2001;119:1717-23.

40. Kaiser L, Wat C, Mills T, Mahoney P, Ward P, Hayden F. Impact of oseltamivir treatment on influenza-related lower respiratory tract complications and hospitalizations. Arch Intern Med 2003;163:1667-72.

41. Monto AS, Webster A, Keene O. Randomized, placebo-controlled studies of inhaled zanamivir in the treatment of influenza A and B: pooled efficacy analysis. J Antimicrob Chemother 1999;44 Suppl B:23-9.

42. Aylward RB, Burdge DR. Ribavirin therapy of adult respiratory syncytial virus pneumonitis. Arch Intern Med 1991;151:2303-4.

43. Englund JA, Piedra PA, Whimbey E. Prevention and treatment of respiratory syncytial virus and parainfluenza viruses in immunocompromised patients. Am J Med 1997;102:61-70

44. McClung HW, Knight V, Gilbert BE, Wilson SZ, Quarles JM, Divine GW. Ribavirin aerosol treatment of influenza B virus infection. JAMA 1983;249:2671-4.

45. Berger A, Drosten Ch, Doerr HW, Sturmer M, Preiser W. Severe acute respiratory syndrome (SARS)--paradigm of an emerging viral infection. J Clin Virol 2004;29:13-22.

46. Manocha S, Walley KR, Russell JA. Severe acute respiratory distress syndrome (SARS): a critical care perspective. Crit Care Med 2003;31:2684-92.

47. Bitar R, Weiser WJ, Avendano M, Derkach P, Low DE, Muradali D. Chest radiographic manifestations of severe acute respiratory syndrome in health care workers: the Toronto experience. AJR Am J Roentgenol. 2004;182:45-8.

48. Chan MS, Chan IY, Fung KH, Poon E, Yam LY, Lau KY. High-resolution CT findings in patients with severe acute respiratory syndrome: a pattern-based approach. AJR Am J Roentgenol 2004;182:49-56.

49. Müller NL, Ooi GC, Khong PL, Zhou LJ, Tsang KW, Nicolaou S. High-resolution CT findings of severe acute respiratory syndrome at presentation and after admission. AJR Am J Roentgenol 2004;182:39-44.

50. Paul NS, Chung T, Konen E, Roberts HC, Rao TN, Gold WL, Mehta S, Tomlinson GA, Boylan CE, Grossman H, Hong HH, Weisbrod GL. Prognostic significance of the radiographic pattern of disease in patients with severe acute respiratory syndrome. AJR Am J Roentgenol 2004;182:493-8.

51. Glezen WP, Greenberg SB, Atmar RL, Piedra PA, Couch RB. Impact of respiratory virus infections on persons with chronic underlying conditions. JAMA 2000;283:499-505.

52. Greenberg SB, Atmar RL. "Chronic airway disease: The infection connection." *In*: Transactions of The American Clinical and Climatological Association 1999;Vol. CX: pp. 38-50.

53. Greenberg SB, Allen M, Wilson J, Atmar RL. Respiratory viral infections in adults with and without chronic obstructive pulmonary disease. Am J Respir Crit Care Med. 2000;162:167-73.

54. Atmar RL, Guy E, Guntupalli KK, Zimmerman JL, Bandi VD, Baxter BD, Greenberg SB. Respiratory tract viral infections in inner-city asthmatic adults. Arch Intern Med 1998;158:2453-9.

55. Tokat O, Kelebek N, Turker G, Kahveci SF, Ozcan B. Intravenous immunoglobulin in adult varicella pneumonia complicated by acute respiratory distress syndrome. J Int Med Res 2001;29:252-5.

56. Triebwasser JH, Harris RE, Bryant RE, Rhoades ER. Varicella pneumonia in adults. Report of seven cases and a review of literature. Medicine (Baltimore) 1967;46:409-23.

57. Tuxen DV, Wilson JW, Cade JF. Prevention of lower respiratory herpes simplex virus infection with acyclovir in patients with the adult respiratory distress syndrome. Am Rev Respir Dis 1987;136:402-5.
58. Hawes S, Seabolt JP. Hantavirus. Clin Lab Sci. 2003;16:39-42.
59. Khan A, Khan AS. Hantaviruses: a tale of two hemispheres. Panminerva Med 2003;45:43-51.
60. Lednicky JA. Hantaviruses. a short review. Arch Pathol Lab Med 2003;127:30-5.
61. Howard MJ, Doyle TJ, Koster FT, Zaki SR, Khan AS, Petersen EA, Peters CJ, Bryan RT. Hantavirus pulmonary syndrome in pregnancy. Clin Infect Dis 1999;29:1538-44.
62. Moolenaar RL, Dalton C, Lipman HB, Umland ET, Gallaher M, Duchin JS, Chapman L, Zaki SR, Ksiazek TG, Rollin PE, et al. Clinical features that differentiate hantavirus pulmonary syndrome from three other acute respiratory illnesses. Clin Infect Dis 1995;21:643-9.
63. Kuwabara S. Guillain-Barre syndrome: epidemiology, pathophysiology and management. Drugs 2004;64:597-610.
64. Solomon T, Willison H. Infectious causes of acute flaccid paralysis. Curr Opin Infect Dis 2003;16:375-81.
65. Jacobs BC, Rothbarth PH, van der Meche FG, Herbrink P, Schmitz PI, de Klerk MA, van Doorn PA. The spectrum of antecedent infections in Guillain-Barre syndrome: a case-control study. Neurology. 1998;51:1110-5.
66. Ray G, Ghosh B, Bhattacharyya R. Acute hepatitis B presenting as Guillain-Barre syndrome. Indian J Gastroenterol 2003;22:228.
67. Steininger C, Popow-Kraupp T, Seiser A, Gueler N, Stanek G, Puchhammer E. Presence of cytomegalovirus in cerebrospinal fluid of patients with Guillain-Barre syndrome. J Infect Dis 2004;189:984-9.
68. Chroni E, Thomopoulos C, Papapetropoulos S, Paschalis C, Karatza CL. A case of relapsing Guillain-Barre syndrome associated with exacerbation of chronic hepatitis B virus hepatitis. J Neurovirol 2003;9:408-10.
69. Laurenti L, Garzia M, Sabatelli M, Piccioni P, Sora' F, Leone G. Guillain-Barre' syndrome following Varicella zoster reactivation in Chronic Lymphocytic Leukemia treated with fludarabine. Haematologica 2002;87:ECR33.
70. Pavone P, Maccarrone F, Sorge A, Piccolo G, Greco F, Caruso P, Sorge G. Guillain-Barre syndrome after varicella zoster virus infections. A case report. Minerva Pediatr 2002;54:259-62.
71. Roccatagliata L, Uccelli A, Murialdo A. Guillain-Barre syndrome after reactivation of varicella-zoster virus. N Engl J Med 2001;344:65-6.
72. Lacaille F, Zylberberg H, Hagege H, Roualdes B, Meyrignac C, Chousterman M, Girot R. Hepatitis C associated with Guillain-Barre syndrome. Liver 1998;18:49-51.
73. da Rosa-Santos OL, Moreira AM, Golfetto CA, Maceira JP, Ramos-e-Silva M. Guillain-Barre syndrome associated with varicella-zoster infection. Int J Dermatol 1996;35:603-4.
74. Ormerod IE, Cockerell OC. Guillain-Barre syndrome after herpes zoster infection: a report of 2 cases. Eur Neurol 1993;33:156-8.
75. Safranek TJ, Lawrence DN, Kurland LT, Culver DH, Wiederholt WC, Hayner NS, Osterholm MT, O'Brien P, Hughes JM. Reassessment of the association between Guillain-Barre syndrome and receipt of swine influenza vaccine in 1976-1977: results of a two-state study. Expert Neurology Group. Am J Epidemiol 1991;133:940-51.

76. Atkins MC, Esmonde TF. Guillain-Barre syndrome associated with rubella. Postgrad Med J 1991;67:375-6.
77. Rabaud C, May T, Hoen B, Maignan M, Gerard A, Canton P. Guillain-Barre syndrome associated with hantavirus infection. Clin Infect Dis 1995;20:477-8.
78. Esack A, Teelucksingh S, Singh N. The Guillain-Barre syndrome following dengue fever. West Indian Med J 1999;48:36-7.
79. Minohara Y, Koitabashi Y, Kato T, Nakajima N, Murakami H, Masaki H, Ishiko H. A case of Guillain-Barre syndrome associated with human parvovirus B19 infection. J Infect 1998;36:327-8.
80. Ahmed S, Libman R, Wesson K, Ahmed F, Einberg K. Guillain-Barre syndrome: An unusual presentation of West Nile virus infection. Neurology 2000;55:144-6.
81. Winer JB. Guillain Barre syndrome. Mol Pathol 2001;54:381-5.
82. Dec GW Jr, Waldman H, Southern J, Fallon JT, Hutter AM Jr, Palacios I. Viral myocarditis mimicking acute myocardial infarction. J Am Coll Cardiol 1992;20:85-9.
83. Maisch B, Ristic AD, Portig I, Pankuweit S. Human viral cardiomyopathy. Front Biosci 2003;8:s39-67.
84. Angelini A, Calzolari V, Calabrese F, Boffa GM, Maddalena F, Chioin R, Thiene G. Myocarditis mimicking acute myocardial infarction: role of endomyocardial biopsy in the differential diagnosis. Heart 2000;84:245-50.
85. Bowles NE, Richardson PJ, Olsen EG, Archard LC. Detection of Coxsackie-B-virus-specific RNA sequences in myocardial biopsy samples from patients with myocarditis and dilated cardiomyopathy. Lancet 1986;1:1120-3.
86. Bultmann BD, Klingel K, Sotlar K, Bock CT, Baba HA, Sauter M, Kandolf R. Fatal parvovirus B19-associated myocarditis clinically mimicking ischemic heart disease: an endothelial cell-mediated disease. Hum Pathol 2003;34:92-5.
87. Galrinho A, Tavares J, Caria R, Veiga C. Myocarditis due to influenza virus complicated by intravascular coagulopathy. Rev Port Cardiol 2000;19:835-8.
88. Greaves K, Oxford JS, Price CP, Clarke GH, Crake T. The prevalence of myocarditis and skeletal muscle injury during acute viral infection in adults: measurement of cardiac troponins I and T in 152 patients with acute influenza infection. Arch Intern Med 2003;163:165-8.
89. Kuhl U, Pauschinger M, Bock T, Klingel K, Schwimmbeck CP, Seeberg B, Krautwurm L, Poller W, Schultheiss HP, Kandolf R. Parvovirus B19 infection mimicking acute myocardial infarction. Circulation 2003;108:945-50.
90. McGregor D, Henderson S. Myocarditis, rhabdomyolysis and myoglobinuric renal failure complicating influenza in a young adult. N Z Med J 1997;110:237.
91. Onitsuka H, Imamura T, Miyamoto N, Shibata Y, Kashiwagi T, Ayabe T, Kawagoe J, Matsuda J, Ishikawa T, Unoki T, Takenaga M, Fukunaga T, Nakagawa S, Koiwaya Y, Eto T.Clinical manifestations of influenza a myocarditis during the influenza epidemic of winter 1998-1999. J Cardiol 2001;37:315-23.
92. Orth T, Herr W, Spahn T, Voigtlander T, Michel D, Mertens T, Mayet WJ, Dippold W, Meyer zum Buschenfelde KH. Human parvovirus B19 infection associated with severe acute perimyocarditis in a 34-year-old man. Eur Heart J 1997;18:524-5.
93. Castleberry JS, Mahon CR. Dengue fever in the Western Hemisphere. Clin Lab Sci 2003;16:34-8.
94. Guzman MG, Kouri G.Dengue and dengue hemorrhagic fever in the Americas: lessons and challenges. J Clin Virol 2003;27:1-13.
95. Halstead SB. Dengue. Curr Opin Infect Dis 2002;15:471-6.

96. Gear JS, Cassel GA, Gear AJ, Trappler B, Clausen L, Meyers AM, Kew MC, Bothwell TH, Sher R, Miller GB, Schneider J, Koornhof HJ, Gomperts ED, Isaacson M, Gear JH. Outbreak of Marburg virus disease in Johannesburg. Br Med J 1975;4:489-93.

97. Heymann DL, Barakamfitiye D, Szczeniowski M, Muyembe-Tamfum JJ, Bele O, Rodier G. Ebola hemorrhagic fever: lessons from Kikwit, Democratic Republic of the Congo. J Infect Dis 1999;179 Suppl 1:S283-6.

98. Richards GA, Murphy S, Jobson R, Mer M, Zinman C, Taylor R, Swanepoel R, Duse A, Sharp G, De La Rey IC, Kassianides C.Unexpected Ebola virus in a tertiary setting: clinical and epidemiologic aspects. Crit Care Med 2000;28:240-4.

99. Srichaikul T, Nimmannitya S. Haematology in dengue and dengue haemorrhagic fever. Baillieres Best Pract Res Clin Haematol 2000;13:261-76.

100. Edelen JS, Bender TR, Chin TD.Encephalopathy and pericarditis during an outbreak of influenza. Am J Epidemiol 1974;100:79-84.

101. Kennedy PG. Viral encephalitis: causes, differential diagnosis, and management. J Neurol Neurosurg Psychiatry 2004;75 Suppl 1:i10-5.

102. McCarthy M. Newer viral encephalitides. Neurologist 2003;9:189-99.

103. Studahl M. Influenza virus and CNS manifestations. J Clin Virol. 2003;28:225-32.

104. Johnson RT. Emerging viral infections of the nervous system. J Neurovirol 2003;9:140-7.

105. Whitley RJ, Cobbs CG, Alford CA Jr, Soong SJ, Hirsch MS, Connor JD, Corey L, Hanley DF, Levin M, Powell DA. Diseases that mimic herpes simplex encephalitis. Diagnosis, presentation, and outcome. NIAD Collaborative Antiviral Study Group. JAMA 1989;262:234-9.

106. Campbell GL, Marfin AA, Lanciotti RS, Gubler DJ. West Nile virus. Lancet Infect Dis 2002;2:519-29.

107. Cushing MM, Brat DJ, Mosunjac MI, Hennigar RA, Jernigan DB, Lanciotti R, Petersen LR, Goldsmith C, Rollin PE, Shieh WJ, Guarner J, Zaki SR. Fatal West Nile virus encephalitis in a renal transplant recipient. Am J Clin Pathol 2004;121:26-31.

108. Centers for Disease Control and Prevention (CDC). First human death associated with raccoon rabies--Virginia, 2003. MMWR Morb Mortal Wkly Rep 2003;52:1102-3.

109. Centers for Disease Control and Prevention (CDC). Human death associated with bat rabies--California, 2003. MMWR Morb Mortal Wkly Rep 2004;53:33-5.

110. Lakeman FD, Koga J, Whitley RJ. Detection of antigen to herpes simplex virus in cerebrospinal fluid from patients with herpes simplex encephalitis. J Infect Dis 1987;155:1172-8.

111. Emig M, Apple DJ. Severe West Nile virus disease in healthy adults. Clin Infect Dis 2004;38:289-92.

112. Petersen LR, Marfin AA. West Nile virus: a primer for the clinician. Ann Intern Med 2002;137:173-9.

113. Roehrig JT, Layton M, Smith P, Campbell GL, Nasci R, Lanciotti RS. The emergence of West Nile virus in North America: ecology, epidemiology, and surveillance. Curr Top Microbiol Immunol 2002;267:223-40.

114. Sampathkumar P. West Nile virus: epidemiology, clinical presentation, diagnosis, and prevention. Mayo Clin Proc 2003;78:1137-43.

115. Jeha LE, Sila CA, Lederman RJ, Prayson RA, Isada CM, Gordon SM. West Nile virus infection: a new acute paralytic illness. Neurology 2003;61:55-9.

116. Kelley TW, Prayson RA, Ruiz AI, Isada CM, Gordon SM. The neuropathology of West Nile virus meningoencephalitis. A report of two cases and review of the literature. Am J Clin Pathol 2003;119:749-53.
117. Solomon T, Willison H. Infectious causes of acute flaccid paralysis. Curr Opin Infect Dis 2003;16:375-81.
118. Jemsek J, Greenberg SB, Taber L, Harvey D, Gershon A, Couch RB. Herpes zoster-associated encephalitis: clinicopathologic report of 12 cases and review of the literature. Medicine (Baltimore) 1983;62:81-97.
119. Nakano A, Yamasaki R, Miyazaki S, Horiuchi N, Kunishige M, Mitsui T. Beneficial effect of steroid pulse therapy on acute viral encephalitis. Eur Neurol 2003;50:225-9.
120. Fagan EA, Williams R. Fulminant viral hepatitis. Br Med Bull 1990;46:462-80.
121. Rezende G, Roque-Afonso AM, Samuel D, Gigou M, Nicand E, Ferre V, Dussaix E, Bismuth H, Feray C. Viral and clinical factors associated with the fulminant course of hepatitis A infection. Hepatology 2003;38:613-8.
122. Tillmann HL, Wedemeyer H, Manns MP. Treatment of hepatitis B in special patient groups: hemodialysis, heart and renal transplant, fulminant hepatitis, hepatitis B virus reactivation. J Hepatol 2003;39:S206-11.
123. Schiodt FV, Davern TJ, Shakil AO, McGuire B, Samuel G, Lee WM. Viral hepatitis-related acute liver failure. Am J Gastroenterol. 2003;98:448-53.
124. Anderson DR, Schwartz J, Hunter NJ, Cottrill C, Bisaccia E, Klainer AS. Varicella hepatitis: a fatal case in a previously healthy, immunocompetent adult. Report of a case, autopsy, and review of the literature. Arch Intern Med 1994;154:2101-6.
125. Anuras S, Summers R. Fulminant herpes simplex hepatitis in an adult: report of a case in renal transplant recipient. Gastroenterology 1976;70:425-8.
126. Kusne S, Schwartz M, Breinig MK, Dummer JS, Lee RE, Selby R, Starzl TE, Simmons RL, Ho M. Herpes simplex virus hepatitis after solid organ transplantation in adults. J Infect Dis 1991;163:1001-7.
127. Harma M, Hockerstedt K, Lautenschlager I. Human herpesvirus-6 and acute liver failure. Transplantation. 2003;76:536-9.
128. Pesik NT, Otten EJ. Severe rhabdomyolysis following a viral illness: a case report and review of the literature. J Emerg Med 1996;14:425-8.
129. Singh U, Scheld WM. Infectious etiologies of rhabdomyolysis: three case reports and review. Clin Infect Dis 1996;22:642-9.
130. Leebeek FW, Baggen MG, Mulder LJ, Dingemans-Dumas AM. Rhabdomyolysis associated with influenza A virus infection. Neth J Med 1995;46:189-92.
131. Seibold S, Merkel F, Weber M, Marx M. Rhabdomyolysis and acute renal failure in an adult with measles virus infection. Nephrol Dial Transplant 1998;13:1829-31.
132. Annerstedt M, Herlitz H, Molne J, Oldfors A, Westberg G. Rhabdomyolysis and acute renal failure associated with influenza virus type A. Scand J Urol Nephrol 1999;33:260-4.
133. Berry L, Braude S. Influenza A infection with rhabdomyolysis and acute renal failure--a potentially fatal complication. Postgrad Med J 1991;67:389-90.
134. Cunningham E, Kohli R, Venuto RC. Influenza-associated myoglobinuric renal failure. JAMA 1979;242:2428-9.
135. Fodili F, van Bommel EF. Severe rhabdomyolysis and acute renal failure following recent Coxsackie B virus infection. Neth J Med 2003;61:177-9.

136. Hollenstein U, Thalhammer F, Burgmann H. Disseminated intravascular coagulation (DIC) and rhabdomyolysis in fulminant varicella infection--case report and review of the literature. Infection 1998;26:306-8.
137. Josselson J, Pula T, Sadler JH. Acute rhabdomyolysis associated with an echovirus 9 infection. Arch Intern Med 1980;140:1671-2.
138. Kantor RJ, Norden CW, Wein TP. Infectious mononucleosis associated with rhabdomyolysis and renal failure. South Med J 1978;71:346-8.
139. Marinella MA. Exertional rhabdomyolysis after recent coxsackie B virus infection. South Med J 1998;91:1057-9.
140. McCabe JL, Duckett S, Kaplan P. Epstein-Barr virus infection complicated by acute rhabdomyolysis. Am J Emerg Med 1988;6:453-5.
141. Morton SE, Mathai M, Byrd RP Jr, Fields CL, Roy TM. Influenza A pneumonia with rhabdomyolysis. South Med J 2001;94:67-9.
142. Nozoe M, Iino T, Nagafuji K, Miyamoto T, Ito H, Gondo H, Harada M. Influenza-induced rhabdomyolysis after autologous peripheral blood stem cell transplantation for malignant lymphoma. Intern Med 2003;42:1127-30.
143. Osamah H, Finkelstein R, Brook JG. Rhabdomyolysis complicating acute Epstein-Barr virus infection. Infection 1995;23:119-20.
144. Pratt RD, Bradley JS, Loubert C, LaRocco A Jr, McNeal RM, Newbury RO, Sawyer MH. Rhabdomyolysis associated with acute varicella infection. Clin Infect Dis 1995;20:450-3.
145. Shenouda A, Hatch FE. Influenza A viral infection associated with acute renal failure. Am J Med 1976;61:697-702.
146. Tanaka T, Takada T, Takagi D, Takeyama N, Kitazawa Y. Acute renal failure due to rhabdomyolysis associated with echovirus 9 infection: a case report and review of literature. Jpn J Med 1989;28:237-42.
147. Ueda K, Robbins DA, Iitaka K, Linnemann CC Jr. Fatal rhabdomyolysis associated with parainfluenza type 3 infection. Hiroshima J Med Sci 1978;27:99-103.
148. Wakabayashi Y, Nakano T, Kikuno T, Ohwada T, Kikawada R. Massive rhabdomyolysis associated with influenza A infection. Intern Med 1994;33:450-3.
149. Wright J, Couchonnal G, Hodges GR. Adenovirus type 21 infection. Occurrence with pneumonia, rhabdomyolysis, and myoglobinuria in an adult. JAMA 1979;241:2420-1.
150. Yamada H, Ono T, Kimura M. Rhabdomyolysis in adults with measles virus infection. Am J Med 2000;108:600.
151. Yoshino M, Suzuki S, Adachi K, Fukayama M, Inamatsu T. High incidence of acute myositis with type A influenza virus infection in the elderly. Intern Med 2000;39:431-2.
152. Alvares-Da-Silva MR, Francisconi CF, Waechter FL. Acute hepatitis C complicated by pancreatitis: another extrahepatic manifestation of hepatitis C virus? J Viral Hepat 2000;7:84-6.
153. Cavallari A, Vivarelli M, D'Errico A, Bellusci R, Scarani P, DeRaffele E, Nardo B, Gozzetti G. Fatal necrotizing pancreatitis caused by hepatitis B virus infection in a liver transplant recipient. J Hepatol 1995;22:685-90.
154. de Oliveira LC, Rezende PB, Ferreira AL, de Freitas AA, de Carvalho AM, Guedes CA, Costa WO. Concurrent acute hepatitis and pancreatitis associated with hepatitis B virus: case report. Pancreas 1998;16:559-61.
155. Ikura Y, Matsuo T, Ogami M, Yamazaki S, Okamura M, Yoshikawa J, Ueda M. Cytomegalovirus associated pancreatitis in a patient with systemic lupus erythematosus. J Rheumatol 2000;27:2715-7.

156. Khanna S, Vij JC. Severe acute pancreatitis due to hepatitis A virus infection in a patient of acute viral hepatitis. Trop Gastroenterol 2003;24:25-6.

157. Majumder AK, Halder A, Talapatra DS, Bhaduri S. Hepatitis E associated with acute pancreatitis with pseudocyst. J Assoc Physicians India 1999;47:1207-8.

158. Shintaku M, Umehara Y, Iwaisako K, Tahara M, Adachi Y. Herpes simplex pancreatitis. Arch Pathol Lab Med 2003;127:231-4.

159. Suskovic T, Vukicevic-Baudoin D, Vucicevic Z, Holjevac I. Severe pancreatitis as first symptom of mumps complicated with pseudocyst and abscess of pancreas. Infection 1997;25:39-40.

160. Yuen MF, Chan TM, Hui CK, Chan AO, Ng IO, Lai CL. Acute pancreatitis complicating acute exacerbation of chronic hepatitis B infection carries a poor prognosis. J Viral Hepat 2001;8:459-64.

161. Crupi RS, Asnis DS, Lee CC, Santucci T, Marino MJ, Flanz BJ. Meeting the challenge of bioterrorism: lessons learned from West Nile virus and anthrax. Am J Emerg Med 2003;21:77-9.

162. Karwa M, Bronzert P, Kvetan V. Bioterrorism and critical care. Crit Care Clin 2003;19:279-313.

163. Han MH, Zunt JR. Bioterrorism and the nervous system. Curr Neurol Neurosci Rep 2003;3:476-82.

164. Henderson DA, Inglesby TV, Bartlett JG, Ascher MS, Eitzen E, Jahrling PB, Hauer J, Layton M, McDade J, Osterholm MT, O'Toole T, Parker G, Perl T, Russell PK, Tonat K. Smallpox as a biological weapon: medical and public health management. Working Group on Civilian Biodefense. JAMA 1999;281:2127-37.

165. Pennington H. Smallpox and bioterrorism. Bull World Health Organ 2003;81:762-7.

166. Breman JG, Henderson DA. Diagnosis and management of smallpox. N Engl J Med 2002;346:1300-8.

167. Nuovo GJ, Plaza JA, Magro C. Rapid diagnosis of smallpox infection and differentiation from its mimics. Diagn Mol Pathol 2003;12:103-7.

168. Baker RO, Bray M, Huggins JW. Potential antiviral therapeutics for smallpox, monkeypox and other orthopoxvirus infections. Antiviral Res 2003;57:13-23.

169. Safrin S, Cherrington J, Jaffe HS. Cidofovir. Review of current and potential clinical uses. Adv Exp Med Biol 1999;458:111-20.

170. Frey SE, Couch RB, Tacket CO, Treanor JJ, Wolff M, Newman FK, Atmar RL, Edelman R, Nolan CM, Belshe RB; National Institute of Allergy and Infectious Diseases Smallpox Vaccine Study Group. Clinical responses to undiluted and diluted smallpox vaccine. N Engl J Med 2002;346:1265-74.

171. Bausch DG, Ksiazek TG. Viral hemorrhagic fevers including hantavirus pulmonary syndrome in the Americas. Clin Lab Med 2002;22:981-1020.

172. Bourdarias B, Perro G, Cutillas M, Castede JC, Lafon ME, Sanchez R. Herpes simplex virus infection in burned patients: epidemiology of 11 cases. Burns 1996;22:287-90.

173. Byers RJ, Hasleton PS, Quigley A, Dennett C, Klapper PE, Cleator GM, Faragher EB. Pulmonary herpes simplex in burns patients. Eur Respir J 1996;9:2313-7.

174. Foley FD, Greenawald KA, Nash G, Pruitt BA Jr. Herpesvirus infection in burned patients. N Engl J Med 1970;282:652-6.

175. Brandt SJ, Tribble CG, Lakeman AD, Hayden FG. Herpes simplex burn wound infections: epidemiology of a case cluster and responses to acyclovir therapy. Surgery 1985;98:338-43.

176. Cherr GS, Meredith JW, Chang M. Herpes simplex virus pneumonia in trauma patients. J Trauma 2000;49:547-9.
177. Fidler PE, Mackool BT, Schoenfeld DA, Malloy M, Schulz JT 3rd, Sheridan RL, Ryan CM. Incidence, outcome, and long-term consequences of herpes simplex virus type 1 reactivation presenting as a facial rash in intubated adult burn patients treated with acyclovir. J Trauma. 2002;53:86-9.
178. Sheridan RL, Schulz JT, Weber JM, Ryan CM, Pasternack MS, Tompkins RG. Cutaneous herpetic infections complicating burns. Burns 2000;26:621-4.
179. Sherry MK, Klainer AS, Wolff M, Gerhard H. Herpetic tracheobronchitis. Ann Intern Med 1988;109:229-33.
180. Jacobs F, Knoop C, Brancart F, Gilot P, Melot C, Byl B, Delforge ML, Estenne M, Liesnard C; Brussels Heart and Lung Transplantation Group. Human herpesvirus-6 infection after lung and heart-lung transplantation: a prospective longitudinal study. Transplantation. 2003;75:1996-2001.
181. Prellner T, Flamholc L, Haidl S, Lindholm K, Widell A. Herpes simplex virus-- the most frequently isolated pathogen in the lungs of patients with severe respiratory distress. Scand J Infect Dis 1992;24:283-92.
182. Ong GM, Lowry K, Mahajan S, Wyatt DE, Simpson C, O'Neill HJ, McCaughey C, Coyle PV. Herpes simplex type 1 shedding is associated with reduced hospital survival in patients receiving assisted ventilation in a tertiary referral intensive care unit. J Med Virol 2004;72:121-5.
183. Bruynseels P, Jorens PG, Demey HE, Goossens H, Pattyn SR, Elseviers MM, Weyler J, Bossaert LL, Mentens Y, Ieven M. Herpes simplex virus in the respiratory tract of critical care patients: a prospective study. Lancet 2003;362:1536-41.
184. Camps K, Jorens PG, Demey HE, Pattyn SR, Ieven M. Clinical significance of herpes simplex virus in the lower respiratory tract of critically ill patients. Eur J Clin Microbiol Infect Dis 2002;21:758-9.
185. Schuller D. Lower respiratory tract reactivation of herpes simplex virus. Comparison of immunocompromised and immunocompetent hosts. Chest. 1994;106:3S-7S.
186. Cook CH, Martin LC, Yenchar JK, Lahm MC, McGuinness B, Davies EA, Ferguson RM. Occult herpes family viral infections are endemic in critically ill surgical patients. Crit Care Med 2003;31:1923-9.
187. Cook CH, Yenchar JK, Kraner TO, Davies EA, Ferguson RM. Occult herpes family viruses may increase mortality in critically ill surgical patients. Am J Surg 1998;176:357-60.
188. Konoplev S, Champlin RE, Giralt S, Ueno NT, Khouri I, Raad I, Rolston K, Jacobson K, Tarrand J, Luna M, Nguyen Q, Whimbey E. Cytomegalovirus pneumonia in adult autologous blood and marrow transplant recipients. Bone Marrow Transplant 2001;27:877-81.
189. Ie S, Rubio ER, Alper B, Szerlip HM. Respiratory complications of pregnancy. Obstet Gynecol Surv. 2002;57:39-46.
190. Lim WS, Macfarlane JT, Colthorpe CL. Treatment of community-acquired lower respiratory tract infections during pregnancy. Am J Respir Med 2003;2:221-33.
191. Atmar RL, Englund JA, Hammill H. Complications of measles during pregnancy. Clin Infect Dis 1992;14:217-26.
192. Maupin RT. Obstetric infectious disease emergencies. Clin Obstet Gynecol. 2002;45:393-404.
193. McKinney WP, Volkert P, Kaufman J. Fatal swine influenza pneumonia during late pregnancy. Arch Intern Med. 1990;150:213-5.

194. Rodrigues J, Niederman MS. Pneumonia complicating pregnancy. Clin Chest Med. 1992;13:679-91.
195. McCarter-Spaulding DE. Varicella infection in pregnancy. J Obstet Gynecol Neonatal Nurs. 2001;30:667-73.
196. Broussard RC, Payne DK, George RB. Treatment with acyclovir of varicella pneumonia in pregnancy. Chest 1991;99:1045-7.
197. Money DM. Antiviral and antiretroviral use in pregnancy. Obstet Gynecol Clin North Am 2003;30:731-49
198. Smego RA Jr, Asperilla MO. Use of acyclovir for varicella pneumonia during pregnancy. Obstet Gynecol 1991;78:1112-6.
199. Chapman SJ. Varicella in pregnancy. Semin Perinatol. 1998;22:339-46.
200. Desselberger U. Herpes simplex virus infection in pregnancy: diagnosis and significance. Intervirology 1998;41:185-90.
201. Frederick DM, Bland D, Gollin Y. Fatal disseminated herpes simplex virus infection in a previously healthy pregnant woman. A case report. J Reprod Med 2002;47:591-6.
202. Kang AH, Graves CR. Herpes simplex hepatitis in pregnancy: a case report and review of the literature. Obstet Gynecol Surv 1999;54:463-8.
203. Baker DA. Issues and management of herpes in pregnancy. Int J Fertil Womens Med. 2002;47:129-35.
204. Godet C, Beby-Defaux A, Agius G, Pourrat O, Robert R.Maternal Herpes simplex virus type 2 encephalitis following Cesarean section. J Infect. 2003;47:174-5.
205. Suarez VR, Hankins GD. Smallpox and pregnancy: from eradicated disease to bioterrorist threat. Obstet Gynecol. 2002;100:87-93.
206. Kojaoghlanian T, Flomenberg P, Horwitz MS. The impact of adenovirus infection on the immunocompromised host. Rev Med Virol. 2003;13:155-71.
207. Meijer E, Boland GJ, Verdonck LF. Prevention of cytomegalovirus disease in recipients of allogeneic stem cell transplants. Clin Microbiol Rev. 2003;16:647-57.
208. Middeldorp JM. Molecular diagnosis of viral infections in renal transplant recipients. Curr Opin Nephrol Hypertens. 2002;11:665-72.
209. Wang WH, Wang HL. Fulminant adenovirus hepatitis following bone marrow transplantation. A case report and brief review of the literature. Arch Pathol Lab Med 2003;127:e246-8.
210. Waser M, Maggiorini M, Luthy A, Laske A, von Segesser L, Mohacsi P, Opravil M, Turina M, Follath F, Gallino A.Infectious complications in 100 consecutive heart transplant recipients. Eur J Clin Microbiol Infect Dis. 1994;13:12-8.
211. Wiesner RH, Rakela J, Ishitani MB, Mulligan DC, Spivey JR, Steers JL, Krom RA. Recent advances in liver transplantation. Mayo Clin Proc. 2003;78:197-210.
212. Nicholson V, Johnson PC.Infectious complications in solid organ transplant recipients. Surg Clin North Am. 1994;74:1223-45.
213. Rubin RH. The direct and indirect effects of infection in liver transplantation: pathogenesis, impact, and clinical management. Curr Clin Top Infect Dis 2002;22:125-54.
214. Alexander BD, Tapson VF. Infectious complications of lung transplantation. Transpl Infect Dis 2001;3:128-37.
215. Cavallo R, Merlino C, Re D, Bollero C, Bergallo M, Lembo D, Musso T, Leonardi G, Segoloni GP, Ponzi AN. B19 virus infection in renal transplant recipients. J Clin Virol 2003;26:361-8.

216. Wendt CH, Hertz MI. Respiratory syncytial virus and parainfluenza virus infections in the immunocompromised host. Semin Respir Infect 1995;10:224-31.

217. Wendt CH, Weisdorf DJ, Jordan MC, Balfour HH Jr, Hertz MI. Parainfluenza virus respiratory infection after bone marrow transplantation. N Engl J Med 1992;326:921-6.

218. Whimbey E, Elting LS, Couch RB, Lo W, Williams L, Champlin RE, Bodey GP. Influenza A virus infections among hospitalized adult bone marrow transplant recipients. Bone Marrow Transplant 1994;13:437-40.

219. Win N, Mitchell D, Pugh S, Russell NH. Successful therapy with ribavirin of late onset respiratory syncytial virus pneumonitis complicating allogeneic bone transplantation. Clin Lab Haematol 1992;14:29-32.

220. Harrington RD, Hooton TM, Hackman RC, Storch GA, Osborne B, Gleaves CA, Benson A, Meyers JD. An outbreak of respiratory syncytial virus in a bone marrow transplant center. J Infect Dis 1992;165:987-93.

221. Lewis VA, Champlin R, Englund J, Couch R, Goodrich JM, Rolston K, Przepiorka D, Mirza NQ, Yousuf HM, Luna M, Bodey GP, Whimbey E. Respiratory disease due to parainfluenza virus in adult bone marrow transplant recipients. Clin Infect Dis 1996;23:1033-7.

222. Morales R, Kirkpatrick M, Browne B, Emovon O. Respiratory syncytial virus pneumonia in an adult renal transplant patient: an unexpected nosocomial infection. Infect Control Hosp Epidemiol 2003;24:548-50.

223. Ohori NP, Michaels MG, Jaffe R, Williams P, Yousem SA. Adenovirus pneumonia in lung transplant recipients. Hum Pathol 1995;26:1073-9.

224. Peigue-Lafeuille H, Gazuy N, Mignot P, Deteix P, Beytout D, Baguet JC. Severe respiratory syncytial virus pneumonia in an adult renal transplant recipient: successful treatment with ribavirin. Scand J Infect Dis 1990;22:87-9.

225. Nichols WG, Corey L, Gooley T, Davis C, Boeckh M. Parainfluenza virus infections after hematopoietic stem cell transplantation: risk factors, response to antiviral therapy, and effect on transplant outcome. Blood 2001;98:573-8.

226. Nichols WG, Gooley T, Boeckh M.Community-acquired respiratory syncytial virus and parainfluenza virus infections after hematopoietic stem cell transplantation: the Fred Hutchinson Cancer Research Center experience. Biol Blood Marrow Transplant. 2001;7 Suppl:11S-15S.

227. Small TN, Casson A, Malak SF, Boulad F, Kiehn TE, Stiles J, Ushay HM, Sepkowitz KA. Respiratory syncytial virus infection following hematopoietic stem cell transplantation. Bone Marrow Transplant 2002;29:321-7.

228. Cane PA, van den Hoogen BG, Chakrabarti S, Fegan CD, Osterhaus AD. Human metapneumovirus in a haematopoietic stem cell transplant recipient with fatal lower respiratory tract disease. Bone Marrow Transplant 2003;31:309-10.

229. Kahn JS. Human metapneumovirus: a newly emerging respiratory pathogen. Curr Opin Infect Dis 2003;16:255-8.

230. Ghosh S, Champlin RE, Englund J, Giralt SA, Rolston K, Raad I, Jacobson K, Neumann J, Ippoliti C, Mallik S, Whimbey E. Respiratory syncytial virus upper respiratory tract illnesses in adult blood and marrow transplant recipients: combination therapy with aerosolized ribavirin and intravenous immunoglobulin. Bone Marrow Transplant 2000;25:751-5.

231. Ljungman P, Gleaves CA, Meyers JD. Respiratory virus infection in immunocompromised patients. Bone Marrow Transplant 1989;4:35-40.

232. Ison MG, Hayden FG. Viral infections in immunocompromised patients: what's new with respiratory viruses? Curr Opin Infect Dis. 2002;15:355-67.

233. Garbino J, Gerbase MW, Wunderli W, Kolarova L, Nicod LP, Rochat T, Kaiser L. Respiratory viruses and severe lower respiratory tract complications in hospitalized patients. Chest. 2004;125:1033-9.

234. Vilchez RA, Dauber J, McCurry K, Iacono A, Kusne S.Parainfluenza virus infection in adult lung transplant recipients: an emergent clinical syndrome with implications on allograft function. Am J Transplant. 2003;3:116-20.

235. Gray J, Wreghitt TG, Pavel P, Smyth RL, Parameshwar J, Stewart S, Cary N, Large S, Wallwork J. Epstein-Barr virus infection in heart and heart-lung transplant recipients: incidence and clinical impact. J Heart Lung Transplant 1995;14:640-6.

236. Parnham AP, Flexman JP, Saker BM, Thatcher GN. Primary varicella in adult renal transplant recipients: a report of three cases plus a review of the literature. Clin Transplant 1995;9:115-8.

237. Singh N. Human herpesviruses-6, -7 and -8 in organ transplant recipients. Clin Microbiol Infect 2000;6:453-9.

238. Taplitz RA, Jordan MC. Pneumonia caused by herpesviruses in recipients of hematopoietic cell transplants. Semin Respir Infect. 2002;17:121-9.

239. Peterson PK, Balfour HH Jr, Marker SC, Fryd DS, Howard RJ, Simmons RL. Cytomegalovirus disease in renal allograft recipients: a prospective study of the clinical features, risk factors and impact on renal transplantation. Medicine (Baltimore) 1980;59:283-300.

240. Sissons JG, Carmichael AJ. Clinical aspects and management of cytomegalovirus infection. J Infect. 2002;44:78-83.

241. Ettinger NA, Bailey TC, Trulock EP, Storch GA, Anderson D, Raab S, Spitznagel EL, Dresler C, Cooper JD. Cytomegalovirus infection and pneumonitis. Impact after isolated lung transplantation. Washington University Lung Transplant Group. Am Rev Respir Dis 1993;147:1017-23.

242. de Maar EF, Verschuuren EA, Harmsen MC, The TH, van Son WJ. Pulmonary involvement during cytomegalovirus infection in immunosuppressed patients. Transpl Infect Dis 2003;5:112-20.

243. Duncan SR, Grgurich WF, Iacono AT, Burckart GJ, Yousem SA, Paradis IL, Williams PA, Johnson BA, Griffith BP. A comparison of ganciclovir and acyclovir to prevent cytomegalovirus after lung transplantation. Am J Respir Crit Care Med 1994;150:146-52.

244. Fishman JA, Doran MT, Volpicelli SA, Cosimi AB, Flood JG, Rubin RH. Dosing of intravenous ganciclovir for the prophylaxis and treatment of cytomegalovirus infection in solid organ transplant recipients. Transplantation 2000;69:389-94.

245. Zamora MR. Controversies in lung transplantation: management of cytomegalovirus infections. J Heart Lung Transplant. 2002;21:841-9.

246. Emanuel D, Cunningham I, Jules-Elysee K, Brochstein JA, Kernan NA, Laver J, Stover D, White DA, Fels A, Polsky B, et al. Cytomegalovirus pneumonia after bone marrow transplantation successfully treated with the combination of ganciclovir and high-dose intravenous immune globulin. Ann Intern Med 1988;109:777-82.

247. Limaye AP. Antiviral resistance in cytomegalovirus: an emerging problem in organ transplant recipients. Semin Respir Infect. 2002;17:265-73.

248. Ramsey PS, Ramin KD. Pneumonia in pregnancy. Obstet Gynecol Clin North Am 2001;28:553-69.

249. Hebart H, Einsele H. Specific infectious complications after stem cell transplantation. Support Care Cancer. 2004 Feb;12(2):80-5.

250. Jenkins FJ, Rowe DT, Rinaldo CR Jr. Herpesvirus infections in organ transplant recipients. Clin Diagn Lab Immunol. 2003;10:1-7.
251. Reusser P. Challenges and options in the management of viral infections after stem cell transplantation. Support Care Cancer. 2002;10:197-203.
252. Hong DS, Jacobson KL, Raad II, de Lima M, Anderlini P, Fuller GN, Ippoliti C, Cool RM, Leeds NE, Narvios A, Han XY, Padula A, Champlin RE, Hosing C. West Nile encephalitis in 2 hematopoietic stem cell transplant recipients: case series and literature review. Clin Infect Dis 2003;37:1044-9.
253. Gouarin S, Vabret A, Gault E, Petitjean J, Regeasse A, de Ligny BH, Freymuth F. Quantitative analysis of HCMV DNA load in whole blood of renal transplant patients using real-time PCR assay. J Clin Virol 2004;29:194-201.

5

TUBERCULOSIS IN THE INTENSIVE CARE UNIT

Charles Feldman
Department of Medicine, University of the Witwatersrand, Johannesburg, South Africa

INTRODUCTION

Tuberculosis is now recognised to be the leading cause of death associated with a single identifiable infectious pathogen in the world (1). According to estimates of the World Health Organisation, which declared tuberculosis to be a global emergency in 1993, there were nearly 2 billion people in the world infected with this micro-organism, with 8 million new cases of active disease and more than 2 million deaths in 1997 (2). 95% of cases of tuberculosis and 98% of tuberculosis deaths occur in developing countries. Tuberculosis has been estimated to cause 7% of all deaths and 26% of preventable deaths in the developing world (1). Most deaths occur in young adults between the ages of 15 and 40 years, during their most economically productive years (2).

Tuberculosis is a disease of the poor and disadvantaged and is therefore concentrated predominantly in the developing world and in poor areas of major cities in the developed world. Although the greatest numbers of cases occur in certain parts of Southeast Asia, the highest incidence of cases is found in sub-Saharan Africa. Nine of the 10 countries with the highest incidence of tuberculosis are in Africa (3). Countries with the highest burden in sub-Saharan Africa include Nigeria, Kenya, Zimbabwe, Tanzania, Uganda, the Democratic Republic of Congo and South Africa (2). In the African region, the estimated incidence is 259 per 100,000 population, compared with 50 per 100,000 in Europe and America (3). This review focuses on severe tuberculosis infection in the developing world.

RESURGENCE OF TUBERCULOSIS

Whereas in the 1960s and 1970s tuberculosis appeared to be decreasing, from the mid-1980s tuberculosis began increasing in incidence worldwide. The reasons for this increase were said to occur in three "epidemics". The first epidemic was the association of tuberculosis with general factors such as poverty, malnutrition, a decrease in socio-economic circumstances, homelessness, decline in tuberculosis control programs, poor compliance with treatment regimens, decreased funding for tuberculosis programs, and civil conflict (2,4).

The second epidemic, occurring particularly in Africa, was the association of tuberculosis with human immuno-deficiency virus (HIV) infection (1,2,4). HIV infection is the greatest risk factor for the progression of latent TB infection to active disease (1). The risk of developing tuberculosis in HIV-seropositive patients is between 3-8% annually, with a 50% lifetime risk (2).

Gachot and co-workers were among the first to describe critical illness in HIV-seropositive patients with tuberculosis, which now appears to be increasing worldwide (5,6). They described 12 cases of severe disseminated tuberculosis infection in patients who were HIV-seropositive, 8 of whom had diffuse pulmonary involvement, which was responsible for the development of acute respiratory failure. Seven of these cases required mechanical ventilation. Seven patients in total died.

The association of HIV infection with tuberculosis in adults has been called the "new tuberculosis" since many of these patients present with unusual or atypical features, The interaction between TB and HIV infection is complex (7,8). It does appear that tuberculosis increases viral replication in HIV-infected individuals and HIV-infected patient who develop tuberculosis appear to have a shortened survival (2).

The third epidemic has been that of multidrug resistant (MDR) tuberculosis, occurring especially in countries such as the United States, particularly in association with intravenous drug abuse, but for which conditions do exist throughout the word (9). A particular concern with drug resistant tuberculosis has been the possibility of greater risk of transmission to other individuals, including nosocomial transmission to health care workers (10). The global extent of drug resistant tuberculosis has not been well defined (11). In some countries, such as the former

Soviet Union, Dominican Republic and Argentina, high rates are seen, but data from most of Africa has not obtained (11,12).

CRITICAL ILLNESS IN ASSOCIATION WITH TUBERCULOSIS

There are a number of conditions that may cause critical illness in patients with tuberculosis (Table 1). Respiratory failure is one of the commonest reasons for intensive care unit admission of patients with tuberculosis. Respiratory failure may be precipitated or complicated by conditions such as pneumothorax, massive haemoptysis, endobronchial obstruction, secondary bacterial infection, and respiratory muscle fatigue (13).

Tuberculosis may remain undiagnosed in patients presenting with unusual features, such as in HIV-seropositive patients and in the setting of respiratory failure (14-20). Delay in diagnosis or failure to diagnose tuberculosis is an important factor responsible for the ongoing mortality of patients, in some cases tuberculosis has only been diagnosed at post-mortem. It therefore remains essential to be aware of the continuing occurrence of tuberculosis in patients and to consider this diagnosis in critically ill cases, especially those presenting in respiratory failure. It may be appropriate in some critically ill cases who are unfit to have further invasive diagnostic testing, or in patients pending the outcome of further investigations, to institute a trial of anti-tuberculosis therapy, with careful follow up (5,6).

Table 1. Major causes of critical illness in patients with tuberculosis
- Acute respiratory failure
- Tuberculous pericarditis
- Tuberculous meningitis
- Tuberculous adrenocortical insufficiency
- Side effects of anti-tuberculosis chemotherapy

Respiratory failure
Respiratory failure in patients with tuberculosis, although relatively uncommon, has been described with regularity over a number of years in both adults and children (13,21-39). The most commonly described entity is that of acute respiratory distress syndrome occurring in association with miliary tuberculosis (13,24-30,32,34,37,39).

The occasional development of an ARDS in patients with tuberculosis is said to depend on the dose and type of bacillary antigen entering the blood stream, as well as on the state of the host's immune response, and particularly the presence of what was previously called delayed hypersensitivity (13,34,37). In miliary tuberculosis (as apposed to localised pulmonary tuberculosis) bacilli and bacillary antigens enter the bloodstream from a large area of the lungs. Whereas small bacillary loads tend to prime a "protective" (predominantly Th1 response), large bacillary loads tend to prime a "destructive" (predominantly Th2 response) (40-42). The factors contributing to the Th1/Th2 shift in tuberculosis infection are being increasingly recognised (40-42).

While ARDS has been described as the cause of respiratory failure in these cases, this diagnosis is often made simply on clinical grounds and is usually not confirmed by either invasive techniques, such as Swan Ganz catheter monitoring, or histology (33). Although disseminated intravascular coagulation (DIC) has been described in a number of patients with miliary tuberculosis, both with and without ARDS (24,25,28,32,34), it appears that this is a complication of the ARDS and not the cause of respiratory failure (13). Nevertheless in miliary tuberculosis it is associated with a poorer prognosis (13,33). The syndrome of antidiuretic hormone secretion (SIADH) has also been said to be a contributory factor to ARDS, by causing increased interstitial and alveolar oedema (23,38).

ARDS has been also described to occur in patients with tuberculous pneumonia and even fibrocavitatory tuberculosis (22,23,31). Some of these cases also had complicating DIC, as has been described in miliary tuberculosis (31). Another cause of respiratory failure in patients with bronchogenic tuberculosis is said to be ventilation/perfusion (V/Q) mismatching due to the presence of large amounts of caseous material in the small airways, with distal collapse of the airways and alveolar spaces, leading to hypoxia and respiratory failure (13).

The mortality of patients with tuberculosis requiring ventilation remains high (37). Where this has been studied, the APACHE II score has consistently predicted a mortality lower than that observed (58), and the mortality in ventilated cases with respiratory failure associated with tuberculosis is higher than that in patients with respiratory failure associated with non-tuberculous pneumonia. Part of this increased mortality may relate to delay in diagnosis of tuberculosis, with subsequent delay in initiation of therapy (37).

There are no formal randomized studies of anti-tuberculosis therapy in critically ill patients on which recommendations for chemotherapy can be based. Nevertheless it usually recommended that 4 standard drugs be given, namely rifampicin, isoniazid, pyrazinamide and ethambutol. Drugs that are available in some countries for parenteral use include rifampicin, isoniazid, and streptomycin, and these may be considered in cases where there are serious concerns of gastrointestinal absorption (13). The use of corticosteroids in patients with tuberculosis is discussed in detail elsewhere (43-45). However, most authors recommend the use of these agents to all patients with acute respiratory failure, provided the patients are initiated on effective anti-tuberculosis therapy (13,32). Similar recommendations are made for patients with HIV-associated tuberculosis in the ICU, particularly if they are extremely ill, severely hypoxic, or have involvement of the meninges or pericardium (6).

Pericarditis

Pericardial involvement by tuberculosis in not uncommon (46). It occurs most commonly due the rupture of a caseous lymph node into the pericardial space. Less often in may occur as part of haematogenous dissemination (13). Occasionally it occurs from contiguous spread from a lung lesion (46). Significant complications of pericardial involvement include the development of cardiac tamponade or constrictive pericarditis (13).

Symptoms are very variable (46). While patients may present with fever and chest pain, the onset may be much more subtle or the presentation be related to the cardiac consequences of the effusion. Cardiac tamponade may occur with either a small rapidly developing effusion or with a large effusion that accumulates slowly over a longer period of time (13). While initially cardiac compression occurs due to fluid alone, later an effusive constrictive process may follow. This has been said to be the most common clinical presentation in South Africa, and is associated with increased pericardial pressure due to due to pericardial effusion in the presence of visceral pericardial constriction (46). This may be a stage in the development of classical constrictive pericarditis (46).

While the diagnosis of the presence of a pericardial effusion is often relatively straightforward on clinical examination and cardiac ultrasound, elucidation of the cause of the effusion (such as being due to tuberculosis) is not easy (13). Fluid obtained from pericardiocentesis is similar to that of pleural fluid – an exudate with a predominance of lymphocytes (13). A positive acid-fast smear is very uncommon and culture is positive in 25-

50% of cases (13). In a study from South Africa, tubercle bacilli were cultured from 59% of 189 pericardial effusions (46). Elevated levels of adenosine deaminase (ADA) in pericardial fluid were found to be helpful in assisting in the diagnosis of tuberculosis in a study from South Africa (46). Pericardial biopsy has also been recommended for diagnosis (13).

Cardiac tamponade due to a pericardial effusion requires urgent therapeutic and diagnostic pericardiocentesis (13). All patients should be treated with anti-tuberculosis drugs as used for pulmonary tuberculosis. A study from South Africa indicated that adjunctive prednisolone reduces the risk of death from pericarditis, reduces the need for repeat pericardiocentesis, but does not change the need for pericardiectomy for constriction (46,47). The addition of prednisone/prednisolone in dosages of 60 – 80 mg/day is therefore usually recommended, unless contraindicated, to be tapered over 6 months (13). Patients not requiring pericardiectomy initially should be followed up with serial echocardiography and managed appropriately. Constrictive pericarditis due to tuberculosis requires anti-tuberculosis therapy, as for pericardial effusion, together with pericardiectomy for persistent constriction (46) Constrictive pericarditis with cardiac decompensation requires an urgent pericardiectomy (13).

Meningitis

Meningeal involvement is probably one of the most serious complications of tuberculosis and the most common cause of death from this infection, particularly in children (48). The increasing incidence of tuberculosis in association with the HIV epidemic appears to be associated with an increased incidence of tuberculous meningitis in adults (49,50). Meningitis occurs as a consequence of the rupture of a subarachnoid focus (Rich's focus that developed following previous haematogenous dissemination (50)), into the subarachnoid space (13). The major impact of the dense inflammatory exudate is on the basal meninges (so-called "basal meningitis")(13, 50). Cranial nerve involvement is quite common and with the inflammatory arteritis, vascular occlusion with cerebral infarction occurs commonly. Blockage of the basal cisterns by the inflammatory exudates may result in obstructive hydrocephalus. Most cases are referred to an intensive care unit for deteriorating neurological status (49).

The diagnosis of tuberculous meningitis is not always straightforward. When typical signs of meningitis are present, the diagnosis may be suspected. However, early clinical signs may be very non-specific (50).

The cornerstone of the diagnosis is the CSF examination, which characteristically shows a raised protein level, a low glucose level, and an increased cell count with a predominance of lymphocytes (13,50). Acid-fast bacilli are seen only in a minority of cases.

Therapy for meningitis, with or without concomitant HIV infection is similar, using the standard drugs. Several of the standard anti-tuberculosis drugs, such as rifampicin, pyrazinamide and isoniazid penetrate reasonably well into the CSF and are recommended as part of therapy. Corticosteroids are often recommended as part of adjunctive therapy, particularly in cases with confusion or altered level of consciousness, focal neurological signs, or hemiplegia (13,50-52). Studies have confirmed the benefit on survival and intellectual outcome of children with tuberculous meningitis, with enhanced resolution of the basal exudates, but no effect on intracranial pressure or incidence of basal ganglion infarction (53).

The mortality rate of tuberculous meningitis remains high and patients are often left with permanent neurological sequelae (49,54,55). The main clinical prognostic features are delay in onset of treatment and neurological status at presentation (49,52).

Adrenocortical Insufficiency
Classic Addison's disease occurs in patients with inactive tuberculosis in whom the adrenal gland tissue has been replaced by granulomas, which are often calcified. In these cases, reactivation of tuberculosis may result in symptoms of both tuberculosis and adrenal insufficiency (13). Sometimes adrenocortical insufficiency does arise as a consequence of miliary tuberculosis in cases in which the adrenal glands are significantly involved (13,56) Rifampicin therapy has been noted to precipitate Addison's disease in patients with borderline adrenal function, as a consequence of increased metabolism of corticosteroids (57). It is important to be aware that a diagnosis of adrenocortical insufficiency may be made even in the presence of apparently normal random levels of cortisol measurements in the blood. Therapy includes hormone replacement, as well as anti-tuberculosis treatment in cases with active infection.

Drug toxicity (57-63)
Adverse drug reactions to anti-tuberculosis drugs are an important cause of additional morbidity, and sometimes mortality, in patients with tuberculosis. It is important to try and differentiate effects due to

disseminated tuberculosis from side effects due to drugs used for treatment. Serious hepatotoxicity is relatively uncommon, but is probably one of the most important adverse drug reactions causing critical illness in patients on treatment, which may be precipitated by several of the commonly used anti-tuberculosis agents. Other potentially serious consequences of drug therapy include effects on the haematological system (thrombocytopenia, neutropenia and aplastic anaemia), the neurological system (seizures), the endocrine system (precipitation of Addison's disease by rifampicin), renal dysfunction, and multiple potentially harmful drug interactions. Anaphylaxis is a very uncommon side effect occasionally described with some of the anti-tuberculosis drugs and drug overdoses with any of the agents may also sometimes be fatal.

THE FUTURE OF TUBERCULOSIS TREATMENT AND PREVENTION

Despite the fact that modern short course treatment is highly effective and cost effective, tuberculosis stills remains a leading cause of disease and suffering (64). It has been nearly 30 years since the introduction of novel compounds for tuberculosis treatment (65). There are a number of compelling reasons why new agents are desperately required, including the need for treatment with fewer drugs, shorter courses, improved MDR control and the treatment of latent infection (65). Deciphering the biology of *Mycobacterium tuberculosis*, through knowledge of its complete genome sequence, enhances this possibility (66). Also of importance is the tuberculosis vaccine, yet despite clear benefits against disseminated childhood infection, its efficacy against adult pulmonary disease has varied widely (67, 68). Developing a novel tuberculosis vaccine represents a daunting task, but is absolutely essential (67, 68).

REFERENCES

1.   Wallis RS, Johnson JL. Adult tuberculosis in the 21st century: pathogenesis, clinical features, and management. Curr Opin Pulm Med 2001; 7: 124-133.

2.   Johnson JL, Ellner JJ. Adult tuberculosis overview: African versus Western perspectives. Curr Opin Pulm Med 2000; 6: 180-186.

3.   Dye C, Scheele S, Dolin P, Pathania V, Raviglione MC for the WHO Global Surveillance and Monitoring Project. Global burden of tuberculosis. Estimated incidence, prevalence and mortality by country. JAMA 1999; 282: 677-686.

4.   Bleed D, Dye C, Raviglione MC. Dynamics and control of the global tuberculosis epidemic. Curr Opin Pulm Med 2000; 6: 174-179.

5.   Gachot B, Wolff M, Clair B, Regnier B. Severe tuberculosis in patients with human immunodeficiency virus infection. Intensive Care Med 1990; 16: 491-493.

6.   Murray JF. HIV-associated tuberculosis: watch for it in your ICU. Intensive Care Med 1990; 16: 487-488.

7.   Del-Amo J, Malin AS, Pozniak A, De Cock KM. Does tuberculosis accelerate the progression of HIV disease? Evidence from basic science and epidemiology. AIDS 1999; 13: 1151-1158.

8.   Whalen C, Horsburgh CR, Hom D, et al. Accelerated course of human immunodeficiency virus infection after tuberculosis. Am J Respir Crit Care Med 1995; 151: 129-135.

9.   Neville K, Bromberg A, Bromberg R, Bonk S, Hanna BA, Rom WN. The third epidemic – multidrug-resistant tuberculosis. Chest 1994; 105: 45-48.

10.   Pearson ML, Jereb JA, Frieden TR, et al. Nosocomial transmission of multidrug-resistant *Mycobacterium tuberculosis*. A risk to patients and health care workers. Ann Intern Med 1992; 117: 191-196.

11.   Willcox PA. Drug-resistant tuberculosis. Curr Opin Pulm Med 2000; 6: 198-202.

12.   Pablos-Mendez A, Raviglione MC, Laszlo A, et al. Global surveillance for anti-tuberculosis-drug resistance, 1994-1997. N Engl J Med 1998; 338: 1641-1649.

13.   Long R. Critical illness due to *Mycobacterium tuberculosis*. In *Principles of Critical Care*. Hall JB, Schmidt GA, Wood LDH, eds. New York: McGraw-Hill, 1992.

14.   Ashba JK, Boyce JM. Undiagnosed tuberculosis in a general hospital. Chest 1972; 61:447-451.

15.   Rosenthal T, Pitlik S, Michaeli D. Fatal Undiagnosed tuberculosis in hospitalised patients. J Infect Dis 1975; 131: S51-S56.

16.   Enarson DA, Grzybowski S, Dorken E. Failure of diagnosis as a factor in tuberculosis mortality. CMA J 1978; 118: 1520-1522.

17.   Bobrowitz ID. Active tuberculosis undiagnosed until autopsy. Am J Med 1982; 72: 650-658.

18.   Katz I, Rosenthal T, Michaeli D. Undiagnosed tuberculosis in hospitalised patients. Chest 1985; 87: 770-774.

19.   Heffner JE, Strange C, Sahn SA. The impact of respiratory failure on the diagnosis of tuberculosis. Arch Intern Med 1988; 148: 1103-1108.

20.   Rieder HL, Kelly GD, Bloch AB. Tuberculosis diagnosed at death in the United States. Chest 1991; 100: 678-681.

21.   Keim LW, Schuldt S, Bedell GN. Tuberculosis in the intensive care unit. Heart and Lung 1977; 6: 624-634.

22.   Agarwal MK, Muthuswamy PP, Benner AS, et al. Respiratory failure in pulmonary tuberculosis. Chest 1977; 72: 605-609.

23.   Sahn SA, Skeff KM. Tuberculous pneumonia with the syndrome of inappropriate secretion of antidiuretic hormone. Cause of the adult respiratory distress syndrome. Chest 1977; 72: 678-681.

24.   Huseby JS, Hudson LD. Miliary tuberculosis and adult respiratory distress syndrome. Ann Intern Med 1976; 85: 609-611.

25.   DeSilva A, Gibson J, Gilbert DN. Miliary tuberculosis and adult respiratory distress syndrome. Ann Intern Med 1977; 86: 659-661.

26.   Hsu JT, Padula JP, Ryan SF. Miliary tuberculosis and respiratory distress syndrome. Ann Intern Med 1978; 89: 140-141.

27.   Raimondi AC, Olmedo G, Roncoroni AJ. Acute miliary tuberculosis presenting as acute respiratory failure. Intensive Care Med 1978; 4: 207-209.

28.   Murray HW, Tuazon CU, Kirmani N, Sheagren JN.   The adult respiratory distress syndrome associated with miliary tuberculosis. Chest 1978; 73: 37-43.

29.   So SY, Yu D. The adult respiratory distress syndrome associated with miliary tuberculosis. Tubercle 1981; 62: 49-53.

30.   Hurwitz SS, Marinopoulos G, Conlan AA, Miller M. Adult respiratory distress syndrome associated with miliary tuberculosis. S Afr Med J 1984; 65: 27-28.

31.   Dyer RA, Potgieter PD.   The adult respiratory distress syndrome and bronchogenic pulmonary tuberculosis. Thorax 1984; 39: 383-387.

32.   Dyer RA, Chappell WA, Potgieter PD. Adult respiratory distress syndrome associated with miliary tuberculosis. Crit Care Med 1985; 13: 12-13.

33.   Levy H, Kallenbach JM, Feldman C, et al. Acute respiratory failure in active tuberculosis. Crit Care Med 1987; 15: 221-225.

34.   Piqueras AR, Marruecos L, Artigas A, Rodriques C. Miliary tuberculosis and adult respiratory distress syndrome. Intensive Care Med 1987; 13: 175-182.

35.   Frame RN, Johnson MC, Eichenhorn MS, et al. Active tuberculosis in the medical intensive care unit: A 15 year retrospective analysis. Crit Care Med 1987; 15: 1012-1014.

36.   Roodt A, Smith C, Feldman C, et al. Apache II severity of illness score in patients with severe active pulmonary tuberculosis. S Afr J Crit Care 1990; 6: 13-16.

37.   Penner C, Roberts D, Kunimoto D, et al. Tuberculosis as a primary cause of respiratory failure requiring mechanical ventilation. Am J Respir Crit Care Med 1995; 151: 867-872.

38.   Vyskocil JJ, Marik P, Greville HW. Survival with tuberculosis pneumonia necessitating mechanical ventilation. Clin Pulm Med 1995; 2: 152-156.

39.   Heyns L, Gie RP, Kling S, et al. Management of children with tuberculosis admitted to a pediatric intensive care unit. Pediatr Infect Dis J 1998; 17: 403-407.

40.   Rook GAW, Hernandez-Pando R. T cell helper types and endocrines in the regulation of tissue-damaging mechanisms in tuberculosis. Immunobiol 1994; 191: 478-492.

41.   Rook GAW, Seah G, Ustianowski A. *M. tuberculosis*: immunology and vaccination. Eur Respir J 2001; 17: 537-557.

42.   Rook GAW, Zumla A. Advances in the immunopathogenesis of pulmonary tuberculosis. Curr Opin Pulm Med 2001; 7: 116-123.

43.   Senderovitz T, Viskum K. Corticosteroids and tuberculosis. Respiratory Medicine 1994; 88: 561-565.

44.   Cunha BA. Pulmonary tuberculosis and steroids. Chest 1995; 107: 1486-1487.

45.   Dooley DP, Carpenter JL, Rademacher S. Adjunctive corticosteroid therapy for tuberculosis: a critical reappraisal of the literature. Clin Infect Dis 1997; 25: 872-877.

46.   Commerford PJ, Strang JIG. Tuberculosis pericarditis. In *A Century of Tuberculosis. South African Perspectives.* Coovadia HM, Benatar SR eds. Capetown: Oxford University Press, 1991.

47.   Strang JIG, Kakaza HHS, Gibson DG, et al. Controlled clinical trial of complete open surgical drainage and of prednisolone in treatment of tuberculosis pericardial effusion in Transkei. Lancet 1998; ii: 759-764.

48.   Moodley M. Neurological tuberculosis. In *A Century of tuberculosis. South African Perspectives.* Coovadia HM, Benatar SR, eds. Capetown: Oxford University Press, 1991.

49.   Verdon R, Chevret S, Laisy J-P, et al. Tuberculous meningitis in adults: review of 48 cases. Clin Infect Dis 1996; 22: 982-988.

50. Berger JR. Tuberculosis meningitis. Curr Opin in Neurology 1994; 7: 191-200.
51. O'Toole RD, Thornton GF, Mukherjee MK, et al. Dexamethasone in tuberculosis meningitis. Relationship of cerebrospinal fluid effects to therapeutic efficacy. Ann Intern Med 1969; 70: 39-49.
52. Kennedy DH, Fallon RJ. Tuberculosis meningitis. JAMA 1979; 241: 264-268.
53. Schoeman JF, van Zyl LE, Laubscher JA, Donald P. Effect of corticosteroids on intracranial pressure, computed tomographic findings, and clinical outcome in young children with tuberculosis meningitis. Pediatrics 1997; 99: 226-231.
54. Weiss W, Flippin HF. The prognosis of tuberculous meningitis in the isoniazid era. Am J Med Sci 1961; 80: 79-86.
55. Fitzsimons JM. Tuberculosis meningitis: A follow-up study on 198 cases. Tubercle 1963; 44: 87-101.
56. Sadler MR, Beresford OD. Miliary tuberculosis associated with Addison's disease. Tubercle 1971; 52: 298-300.
57. Zent C. Toxicity of anti-tuberculous medication. SA J Epidemiol Infect 1994; 9: 5-9.
58. Keven K, Uysal AR, Erdogan G. Adrenal function during tuberculous infection and effects of antituberculosis treatment on endogenous and exogenous steroids. Int J Tubercle Lung Dis 1998; 2: 419-424.
59. Rossouw JE, Saunders SJ. Hepatic complications of antituberculous therapy. Q J Med 1975; 173: 1-16.
60. Zierski M, Bek E. Side-effects of drug regimens used in short-course chemotherapy for pulmonary tuberculosis: a controlled clinical study. Tubercle 1980; 61: 41-49.
61. Thompson NP, Caplin ME, Hamilton MI, et al. Anti-tuberculosis medication and the liver: dangers and recommendations in management. Eur Respir J 1995; 8: 1384-1388.
62. Schaberg T. The dark side of antituberculosis therapy: adverse events involving liver function. Eur Respir Rev 1995; 4: 1247-1249.
63. Mitchell I, Wendon J, Fitt S, Williams R. Anti-tuberculous therapy and acute liver failure. Lancet 1995; 345: 555-556.
64. Grange JM, Zumla A. Advances in the management of tuberculosis: clinical trials and beyond. Curr Opin Pulm Med 2000; 6: 193-197.
65. O'Brien RJ, Nunn PP. The need for new drugs against tuberculosis. Obstacles, opportunities, next steps. Am J Respir Crit Care Med 2001; 162: 1055-1058.
66. Cole ST, Brosch R, Parkhill J, et al. Deciphering the biology of *Mycobacterium tuberculosis* from the complete genome sequence. Nature 1998; 393: 537-544.
67. Malin A, Young DB. Designing a vaccine for tuberculosis. Unravelling the tuberculosis genome – can we build a better BCG? Br Med J 1996; 312: 1495.
68. Doherty TM, Andersen P. Tuberculosis vaccine development. Curr Opin Pulm Med 2002; 8: 183-187.

6

Tuberculosis in the Intensive Care Unit: The North American Perspective

Loren C. Denlinger and Jeffrey Glassroth
Division of Pulmonary & Critical Care, Department of Medicine, University of Wisconsin-Madison and the University of Wisconsin Hospital & Clinics, Madison, Wisconsin, USA

1   INTRODUCTION AND EPIDEMIOLOGY OF TUBERCULOSIS

There are many similarities between the recognition and management of patients with tuberculosis in the intensive care unit and in the outpatient clinic. However, there are also some key differences. This chapter will review special presentations of tuberculosis in critically ill patients, unique management issues in this population, and practical aspects of infection control.

Global estimates suggest that one third of the world's population has been infected by *Mycobacterium tuberculosis*. Ten percent of infected individuals will progress to active disease, at least half of them within the first two years following infection. All together, there are an estimated eight to sixteen million new cases a year with two to three million deaths. Less than 10% of these patients have coincident human immunodeficiency viral infection, however, when this occurs the case fatality rate is over 50% and progression from infection to active disease occurs at a rate of at least 8% per year. Whereas the greatest number cases remain concentrated in Southeast Asia, sub-Saharan Africa and eastern Europe, recent trends in global travel predict resurgence of disease outside of these regions (1).

After a thirty-year decline in the number of cases, there was an increase in the incidence within the United States between 1985 and 1992, followed by another decline through 2002 (2-4). Although all demographic groups were affected by this increase, over half of the cases in the United States were immigrants, especially women from Asia or Central America. Enhanced transmission may have also contributed, reflected by a substantial rise in the incidence of cases in children four years of age or younger (2). Co-infection with HIV, as well as multi-drug resistance, also accounted for a large percentage of the increase with enhanced morbidity, particularly in Caucasian and African American men (4, 5). As patients with compromised immune systems are now living longer, it stands to reason that the incidence of tuberculosis in the ICU will also increase.

2   FORMS OF TUBERCULOSIS PERTINENT TO THE INTENSIVIST

Delayed recognition of tuberculous disease in hospital patients not only contributes to nosocomial spread, but also has been associated with an increased risk of mortality in hospitalized patients (6, 7). Nearly three quarters of hospitalized patients with active TB have overall management delays greater than 24 hours, and this is both with respect to suspicion of the disease, as well as its diagnosis and the initiation of treatment (8). Risk factors for a delayed diagnosis include age greater than 65 years, a lack of respiratory symptoms, the absence of hemoptysis or cavitary lung disease, and a misinterpretation of "unusual" chest radiographs (8, 9). Atypical pulmonary or extrapulmonary TB is associated with HIV co-infection, and recently with the use of infliximab or other immunomodulatory agents directed against tumor necrosis factor-α, possibly due to reductions in macrophage apoptosis necessary for mycobacterial killing within granulomas (10).

The identification of radiographic patterns of tuberculous disease can be a key to early recognition. Primary disease manifestations include unilateral hilar adenopathy with lower lobe infiltrates and/or pleural effusions (11). The classic distributions for reactivated tuberculosis include infiltration with or without cavitations of the apical and posterior segments of the upper lobes, followed by the superior segments of the lower lobes, and/or other segments in combination with upper lobe involvement (11). Other radiographic presentations tend to occur in patients with comorbidities such as cancer, diabetes mellitus and alcohol abuse (11, 12). These include exclusive infiltration of the middle or lower lobes, miliary TB, and tuberculomas (11). HIV co-infected persons, particularly those with advanced immune compromise, are particularly prone to such presentations and may even have "normal" chest radiographs despite active pulmonary disease. Unfortunately from a diagnostic perspective, lower lobe presentations tend to be smear negative (11). The following sections will briefly provide information that, in combination with radiographic findings and microbiologic studies, may assist in making the diagnosis of tuberculosis in patients admitted to the ICU.

2.1   Respiratory Failure/ARDS

Although uncommon, both bronchopulmonary and miliary tuberculosis can cause substantial hypoxemia (13-17), and the occurrence of respiratory failure has been associated with delay in microbiologic diagnosis (18). Almost all of the reported cases had a risk factor for developing active disease, including ethanol abuse, malnutrition, diabetes mellitus, or chronic use of corticosteroids (14-18). Fever, cough, and dyspnea were uniformly present for at least a week prior to presentation and often much longer, which is in contrast to the usual rapid course of prodromal illness for nontuberculous bacterial or viral causes of respiratory failure (14). With this type of subacute presentation, common signs or laboratory findings that may help with the differential diagnosis in favor of tuberculosis include hepatomegally, anemia, hypoalbuminemia, and consumptive coagulopathy (13-16, 18).

The incidence of respiratory failure leading to mechanical ventilation in patients with established miliary disease or tuberculous pneumonia is estimated at 18.9 and 0.8% respectively (17). This progression in both cases appears to be tightly linked to the onset of the adult respiratory distress syndrome (ARDS), suggesting that patients with bronchopulmonary disease are relatively spared from ARDS possibly due to the localized concentration of mycobacterial antigens (17). After adjusting for severity of illness, the hospital mortality for all patients with tuberculosis requiring mechanical ventilation was similar to patients with ARDS from any cause (69 vs. 56%) but higher than the mortality from nontuberculous bacterial pneumonia requiring mechanical ventilation but without ARDS (36%) (17). The reasons postulated for this increased mortality include a delay in appropriate anti-microbial therapy, difficulties associated with the administration and tolerance of anti-tuberculosis drugs, and the frequent lack of quick improvement even in the setting of appropriate therapy.

2.2   Septic Shock

Whereas hypotension commonly complicates respiratory failure due to Gram-negative pathogens, septic shock is felt to be uncommon ($\leq$7%) in patients with tuberculous ARDS (16). Indeed, it is seen almost exclusively in immunocompromised patients with miliary disease (19-21), occasionally with a clear chest film on admission (19, 22). Tuberculous septic shock has also been described as a complication of prior therapy with prednisone for co-infection with *Pneumocystis carinii* in the setting of HIV (20). As mycobacterial blood cultures using the lysis-centrifugation method have recently been shown to be at least as sensitive as bone marrow biopsies for diagnosing disseminated tuberculosis (23), these should be considered for any immunocompromised patient presenting with septic shock.

2.3   Adrenal Insufficiency

Prior to the availability of antibiotics, tuberculosis was responsible for nearly 70% of all cases of Addison's disease (24). However, in a recent retrospective case series, granulomas were only found in 52 of 871 autopsies from patients with active tuberculosis, yielding an overall incidence of adrenal involvement of 6% of all patients with TB (25). An additional 3 cases were detected among surviving patients undergoing 270 adrenalectomies for indications without suspicion of TB. In this series, thirty percent of the patients had extrapulmonary disease, and the adrenal gland was the fifth most common site, behind liver, spleen, kidney and bone (25). Interestingly, only seven of the 55 patients with adrenal involvement presented with signs and symptoms of Addison's disease (25). This is consistent with biochemical studies revealing subnormal cortisol responses to administration of cosyntropin in only 8 of 88 patients with active disease prior to starting therapy and in only one of these 88 patients after four weeks antibiotic therapy (26). Nonetheless, given the prevalence of relative adrenal insufficiency in ICU patients (27), this entity must be considered in any patient with tuberculosis presenting with septic shock and replacement of stress-dose steroids may be beneficial (28). Conversely, tuberculosis must be considered as a potential explanation for patients presenting with adrenal insufficiency.

2.4   Massive Hemoptysis

Tuberculosis is still felt to be a major etiologic consideration in patients with massive hemoptysis, defined as anywhere from 100 mL of blood in 24 hr to 1000 mL in several days, with an associated acute mortality of 7 to 32 percent (29, 30). Often, this is in the setting of active disease with a high burden of organisms. It can occur either with bronchiolar ulceration and necrosis of adjacent bronchial vessels during acute disease, or by rupture of a Rasmussen's aneurysm from the pulmonary arterial circulation eroding through a thick walled cavity (30). Additionally, massive hemoptysis can occur after resolution of active disease by the broncholithic passage of a calcified lymph node (30). Early surgical intervention has been recommended for definitive treatment, with few studies to document overall efficacy in this high-risk population (29). Securing the airway, often by the use of a double-lumen endotracheal tube, is essential followed by bronchoscopic localization of the source of bleeding as the diagnostic procedure of choice (30) and by blood replacement. Temporizing measures including iced saline irrigation, topical epinephrine or fibrinogen, balloon tamponade, and /or vessel embolization appear to be associated with variable success (30). Unfortunately, and despite these measures, bleeding can be massive and exsanguination may occur.

2.5   Other Presentations

In the last few decades, the number of cases of extrapulmonary tuberculosis has been relatively fixed at roughly 4000 per year in the U.S., and some of these cases have been accompanied by pulmonary involvement (31, 32). However, the incidence of extrapulmonary disease appears to be increasing (17.5% of all tuberculous cases in 1986 vs. historical values of about 11 to 13%), reflective of a decline in pulmonary cases and a rise in the number of immunocompromised patients particularly with HIV co-infection (31, 32). The sites of infection for these cases are listed in decreasing order of prevalence; lymphatic (5.4% of the total), pleural (4.0%), genitourinary (2.1%), bone and joint (1.7%), other (1.6%), miliary (1.3%), meningeal (0.8%), and peritoneal (0.6%) (32). Unfortunately, these data have not been stratified according to the level of disease severity (i.e. ICU vs. hospital ward vs. community). Nonetheless, selected extrapulmonary presentations will be discussed here as they pertain to the potential contribution to nosocomial spread of tuberculosis within an ICU or to an increased acute mortality rate associated with delayed diagnosis.

Patients with either genitourinary, lymph node or dermatologic forms of tuberculosis tend to present to the intensive care unit for unrelated reasons and may or may not have pulmonary involvement. Both urinary TB and scrofula have a similar pathogenesis involving the erosion of preexisting granulomas through local tissue planes (33, 34). Although the potential exists for limited environmental shedding of viable organisms, this routinely is without aerosol formation such that nosocomial spread is very unlikely by this route, unless there is high velocity irrigation of diseased bone, joint or skin. Urinary TB is suggested by unexplained (i.e. "sterile") pyuria, particularly in the presence of renal calyx dilatation, focal calcium deposits, cavity formation, or multiple segments of partial ureteral obstruction (33). Lymph node tuberculosis is suggested by the presence of an indurated, cool subcutaneous nodule over the neck, clavicle, knees or ankles with a sinus tract draining caseous material, and is often a reflection underlying soft tissue or bone/joint involvement, necessitating radiographic investigation of these areas (34). Other dermatologic hallmarks of tuberculous reactivation with or without pulmonary involvement include lupus vulgaris (coalescent red-brown papules forming a plaque with raised edges and scarred centers on the head and neck), and less commonly tuberculosis cutis orificialis (nonhealing periorificial ulcers in immunocompromised patients). Fortunately in all of these cases, acid-fast stains of spun urine samples, drainage or biopsy material, are very sensitive and can facilitate diagnosis, and offer an explanation for pulmonary infiltrates when these are present.

Although uncommon, tuberculous pericarditis and meningitis have a high mortality rate, and are more likely than other presentations to escape diagnosis until the time of autopsy (35-37). Whereas pericardial involvement almost always reflects the presence of clinically silent tuberculosis at other sites (35), the meningeal form often presents with a normal chest radiograph and occasionally represents the only site of involvement in a previously healthy patient without identifiable risk factors (36). Unfortunately, the clinical presentations for these two extrapulmonary manifestations do not distinguish them from non-tuberculous etiologies. Additionally, the sensitivity of acid-fast smears of pericardial and cerebral spinal fluid are low. However, tuberculosis must be in the differential diagnosis for patients presenting to the ICU with tamponade, particularly immunocompromised patients or those from select racial/ethnic groups or communities with a relatively high rate of TB who also have a bloody, lymphocytic and/or monocytic exudative effusion (35). Similarly, tuberculosis must be strongly considered for critically ill patients with mental status changes, signs and symptoms consistent with meningitis, who also have a lymphocytic pleocytosis with a glucose less than 40 in their cerebral spinal fluid with a negative Gram stain and cryptococcal antigen. Adenosine deaminase (ADA) analysis and/or the polymerase chain reaction (PCR) coupled with gene probing of this fluid may help speed diagnosis (see below). Determining the level of interferon-γ (IFN-γ) in these fluids may also be useful in diagnosis. Finally, due to the intensity of the inflammatory and fibrotic reactions at these two sites, adjunctive corticosteroids (in addition to standard therapy) are recommended in the management of patients with either tuberculous pericarditis or meningitis (38).

3   ADVANCES IN DIAGNOSIS

Submission of clinical specimens for acid fast staining and culture remains the gold standard for diagnosis, particularly with respect to drug susceptibility determination for the isolate (39). Unfortunately, the sensitivity of microscopic examination is highly variable depending on the source of the specimen. For example, although the culture sensitivity and specificity for sputum samples from patients with active pulmonary disease is 85 and 98% respectively, up to fifty percent of the samples can be smear negative, necessitating submission of multiple specimens and potentially contributing to diagnostic delay (39). Analysis of adenosine deaminase levels from a variety of fluids has high specificity, although sometimes lacks sensitivity (40). Additionally, the level of IFN-γ has recently received attention, particularly in making the diagnosis of tuberculous pleural effusion (41). Thus, there has been intense effort in the development of rapid diagnostic tests.

Two areas of advance include tests to better recognize latent infection (see the infection control section below for expanded discussion), as well as those directed at diagnosing active disease in the setting of negative acid fast smears. With respect to the former, there are a variety of assays to measure the production of gamma interferon, or the cells producing it, in response to stimulation with *M. tuberculosis*-specific antigens (42-44). These have been validated in patients at high risk for latent infection with reasonable sensitivities, however, they are not yet widely available and have not been approved for patients with active disease. By contrast, the U.S. Food and Drug Administration (FDA) has approved the Enhanced MTD test for the diagnosis of active disease using respiratory specimens. This oligonucleotide amplification test has recently been validated in 338 patients clinically stratified as to whether they had low, intermediate, or high pretest probabilities of active disease, according to evaluation of their symptoms, risk factors, tuberculin skin test results, and plain chest films (45). With excellent positive and negative predictive values (59%/100%; 100%/95%; 91%/51%) in the three pretest groups respectively, this test clearly outperformed microscopic examination (ppv 36%, 30%, 90%; npv 96%, 71%, 37%) (45). Other PCR based assays exist, some of which are also being tested with extrapulmonary samples (46-48). Thus, confirmation of positive smear results, or probing smear negative samples from high risk patients can now be rapidly and reliably performed, minimizing the delay to diagnosis and initiation of treatment.

Finally, there have also been advances in the methods used for drug susceptibility testing. The Mycobacteria Growth Inhibition Test (MGIT) BACTEC system is fully automated, and does not require radiometric methods standard to conventional testing protocols. The performance of this test has been good. However, there are at least two reports suggesting that it may be prone to contamination (49, 50) raising a cautionary note prior to wide spread implentation.

4   TREATMENT AND MANAGEMENT ADVICE

The treatment of patients with active tuberculosis with or without HIV coinfection has recently been reviewed (51, 52), and extensive guidelines have been updated by a joint commission of the American Thoracic Society, the Center for Disease Control and Prevention, and the Infectious Disease Society of America (38). Briefly, there are now ten antibiotics with FDA approval (and several with anti-tuberculous activity but not formal approval) for the use against *M. tuberculosis*. Despite this repertoire, the preferred initial regimen remains isoniazid, rifampin, ethambutol, and pyrazinamide. Whereas isoniazid has the most killing activity against rapidly dividing mycobacteria, rifampin or other rifamycins have become the cornerstone of tuberculous therapy because of their sterilizing ability in dormant populations. Pyrazinamide has preferential activity in acidic environments and is therefore useful cavitary disease. Ethambutol helps maintain susceptibility to rifamycins in the setting of pre-existing isoniazid resistance, and may add to a regimens early mycobactericidal activity. Other first-line drugs include rifapentine and rifabutin (not FDA approved). Second-line agents include cycloserine, ethionamide, streptomycin, amikacin/kanamycin, capreomycin, p-aminosalicylic acid, ciprofloxacin (or other fluoroquinolones) and clofazamine. The initial phase of treatment should include four drugs for at least two months, followed by a continuation phase with two drugs (preferably isoniazid and rifampin) determined by culture susceptibilies and the site and extent of disease burden (38).

Although well-accepted, reduced-frequency (i.e. intermittent) regimens exist for both the initial and continuation phases, critically ill patients should receive daily therapy due to the risk of altered pharmacokinetics in this population. Because ethambutol and pyrazinamide are only available for oral administration, every effort should be made to initiate early feeding and drug delivery via a Dobhoff or similar tube, even within twelve hours of major abdominal surgery and in advance of the return of bowel sounds and flatus. If the oral route cannot be tolerated due to severe gastroparesis, vomiting, or diarrhea with malabsorption, parenteral therapy can be initiated with isoniazid, rifampin, streptomycin and a fluoroquinolone, followed by a transition to a standard regimen when bowel function is restored. Other factors influencing pharmacokinetics include the presence of renal dysfunction, necessitating dose adjustments for most agents. Moreover, patients with severe liver dysfunction at baseline should not be given a standard regimen due to the combination of multiple potentially hepatotoxic agents. In this case, a preferable regimen might be rifampin, ethambutol, and a fluoroquinolone or aminoglycoside (38). If there are major pharacokinetic concerns about an individual critically ill patient, it is possible and perhaps desirable to monitor drug levels (53), however, these targets are not well established making this area of therapy controversial (38). Finally, many of these antibiotics have significant drug interactions with agents commonly used in the ICU, including anticoagulants, anticonvulsants, cardiovascular agents, psychotropic drugs, and steroids or other drugs used in organ transplantation. Thus, clinical vigilance is needed, often supplemented by therapeutic drug monitoring of the non-tuberculous agents (38).

5 CONTROLLING TB TRANSMISSION IN THE INTENSIVE CARE UNIT

Effective tuberculous infection control requires a thorough understanding of the transmission of this organism. The acts of breathing, coughing, sneezing, speaking and singing normally cause the release of droplets of varying size, which also carry tubercle bacilli from an individual with active pulmonary disease. The largest of these droplets rapidly settle onto surfaces, which are contaminated but generally not considered infectious from a pulmonary standpoint. Evaporation of medium particles causes the formation of droplet nuclei roughly 1 to 5 μm in size. Animal experiments have shown that nuclei of this size are required to directly infect the alveolus, and that particle deposition in the larger airways does not render a productive infection. These nuclei remain airborne and infectious for several hours (54).

Preventing nosocomial spread of TB requires attention to all of the following; 1) clinical suspicion, 2) prospective isolation of high risk patients prior to establishing diagnosis, 3) engineering controls of ambient air to prevent dissemination of aerosols from an infected patient's room, 4) personnel protection in the form of masks, 5) rendering infectious patients non-infectious by the initiation of effective therapy, and 6) administrative policies and programs to follow up with and initiate treatment of inadvertently exposed personnel (54, 55). Unfortunately, attempts to use clinical risk factors to reduce the numbers of patients with non-tuberculous disease requiring respiratory isolation during their workup have unacceptably high failure rates (20% or more) (56, 57). Thus, many institutions have adopted policies such that the decision to send sputum for AFB analysis on any hospitalized patient mandates respiratory isolation until three smears are negative.

From a practical standpoint, the riskiest patients to the ICU personnel are any patients with respiratory symptoms admitted prior to the suspicion of tuberculosis. Initiation of mechanical ventilation dramatically reduces the risk of transmission of aerosols because it is a closed system and the exhalation port is filtered. However, this does not eliminate the need for isolation and personnel masking due to unexpected breeches that can occur. As discussed above, patients exclusively with extrapulmonary tuberculosis are much less contagious due to the lack of aerosol formation. Unless concomitant pulmonary involvement can be excluded by negative sputum AFB smears, these patients should also be isolated.

Unfortunately, inadvertent exposure of ICU personnel can occur prior to suspicion, isolation, and implementation of respiratory protection. Any exposed personnel should have a PPD test within the first week following exposure to document prior exposure history, followed by a second PPD test three months later. With documented exposure to a patient with active disease, the "cut point" for a positive result becomes 5 mm of induration, reduced from the 10 mm point normally used for health care personnel so as to increase the sensitivity of the test. According to recently revised guidelines, a positive PPD test is an indication to consider treatment to prevent the occurrence of latent disease, regardless of age. This recommendation reflects a lower incidence of isoniazid-related hepatitis than previously documented, in combination with the recent resurgence in TB incidence. If the isolate from the index patient is susceptible, isoniazid (5 mg/kg PO daily, max 300 mg/day) remains the standard for a course of therapy for nine months. Shorter regimens may be effective including the combined use of rifampin and pyrizinamide, however, the latter may be associated with increased risk of liver toxicity. Successful completion of a prophylactic regimen reduces the risk of development of active disease by 93% (58).

6   A LOOK TO THE FUTURE

The ultimate goal in TB research is the development of a safe, effective, and durable vaccine for the prevention of primary infection, and possibly therapeutic benefit in the setting of active disease. Several candidate vaccines and immunization strategies are under investigation (59), however, a major limitation is that the production of gamma interferon, the main correlate of protective immunity in animal models, does not predict mycobacterial immunity in humans (60). Thus, at least five candidate vaccines are being tested in large Phase 3 clinical trials without the benefit of preliminary surrogate data from Phase 2 studies (59).

Genomic technology and information will also have a large impact on the management of this disease. Genotype information from mycobacterial isolates is already being used for epidemiological research, as well as for the identification of new antigens to be used as targets for subunit vaccines (61). Additionally, efforts are underway to identify genetic factors that contribute to variability in the immune response to tuberculous infection in humans. Several candidate genes, including gamma interferon, interleukin-10, and the natural resistance-associated macrophage protein 1 (NRAMP-1), have polymorphisms that are associated with an increased risk of active disease in select populations (62, 63). A genome-wide linkage analysis also suggests that there may be X-linked gene polymorphisms associated with susceptibility in Africans (64).

Unfortunately, genetic association studies have a high false positive rate, such that the ability to apply human genomic data to clinical practice will require careful evaluation of its clinical validity and utility. Making the transition from the epidemiological identification of a genetic risk factor to a surrogate endpoint for clinical outcome can be greatly facilitated when there is a reproducible biochemical assay that correlates with both the genetic changes and the outcome surrogate. An example of this may be the nucleotide receptor, $P2X_7$. Polymorphisms of this gene have been associated with protection from smear-positive disease in a Gambian population (65). This gene encodes an ion channel and pore expressed by most classes of leukocytes, and is involved with phagolysosomal maturation and killing of intracellular mycobacteria (66). Individuals with other loss-of-function genetic defects in this $P2X_7$ pore have monocytes that are less able kill the BCG strain *in vitro* (67). This $P2X_7$ pore is defective in roughly 15% of healthy adults and can be screened rapidly (68), thus providing the foundation for future clinical trials evaluating the risk of reactivation from latent disease. Collectively, genomic advances may allow for rapid diagnosis of tuberculosis in critically ill patients and could even direct the administration of immunomodulatory adjunctive therapy.

Acknowledgements

Dr. Denlinger is supported by the Will Rogers Research Institute
Dr. Glassroth is the George R. and Elaine Love Professor and Chair, Department of Medicine.

REFERENCES

1. Dye, C., S. Scheele, P. Dolin, V. Pathania, and M. C. Raviglione. 1999. Global burden of tuberculosis: estimated incitdence, prevalence, and mortality by country. JAMA 282:677-86.
2. Cantwell, M. F., D. E. Snider, G. M. Cauthen, and I. M. Onorato. 1994. Epidemiology of tubeculosis in the United States, 1985 through 1992. JAMA 272:535-9.
3. McCray, E., C. M. Weinbaum, C. R. Braden, and I. M. Onorato. 1997. The epidemiology of tuberculosis in the United States. Clin Chest Med 18:99-113.
4. Iademarco, M. F., and K. G. Castro. 2003. Epidemiology of tuberculosis. Semin Respir Infect 18:225-40.
5. Hopewell, P. C. 1992. Impact of human immunodeficiency virus infection on the epidemiology, clinical features, management and control of tuberculosis. Clin Infect Dis 15:540-7.
6. Pablos-Mendez, A., T. R. Sterling, and T. R. Frieden. 1996. The relationship between delayed or incomplete treatment and all cause mortality in patients with tuberculosis. JAMA 276:1223-8.
7. Sacks, L. V., and S. Pendle. 1998. Factors related to in-hospital deaths in patients with tuberculosis. Arch Intern Med 158:1916-22.
8. Rao, V. K., E. P. Iademarco, V. J. Fraser, and M. H. Kollef. 1999. Delays in the suspicion and treatment of tuberculosis among hospitalized patients. Ann Intern Med 130:404-11.
9. Mathur, P., L. Sacks, G. Auten, R. Sall, C. Levy, and F. Gordin. 1994. Delayed diagnosis of pulmonary tuberculosis in city hospitals. Arch Intern Med 154:306-10.
10. Keane, J., S. Gershon, R. P. Wise, E. Mirabile-Levens, J. Kasnica, W. D. Schwieterman, J. N. Siegel, and M. M. Braun. 2001. Tuberculosis associated with infliximab, a tumor necrosis factor α-neutralizing agent. New England J Med 345:1098-104.
11. Khan, M. A., D. M. Kovnat, B. Bachus, M. E. Whitcomb, J. S. Brody, and G. L. Snider. 1977. Clinical and roentgenographic spectrum of pulmonary tuberculosis in the adult. Am J Med 62:31-6.
12. Perez-Guzman, C., A. Torres-Cruz, H. Villarreal-Velarde, and M. H. Vargas. 2000. Progressive age-related changes in pulmonary tuberculosis images and the effect of diabetes. Am J Respir Crit Care Med 162:1738-40.

13. Agarwal, M. K., P. P. Muthuswamy, R. S. Shah, and W. W. Addington. 1977. Respiratory failure in pulmonary tuberculosis. Chest 72:605-9.

14. Murray, H. W., C. U. Tuazon, N. Kirmani, and J. N. Sheagren. 1978. The adult respiratory distress syndrome associated with miliary tuberculosis. Chest 73:37-43.

15. Dyer, R. A., and P. D. Potgieter. 1984. The adult respiratory distress syndrome and bronchogenic pulmonary tuberculosis. Thorax 39:383-7.

16. Dyer, R. A., W. A. Chappell, and P. D. Potgieter. 1985. Adult respiratory distress syndrome associated with miliary tuberculosis. Crit Care Med 13:12-5.

17. Penner, C., D. Roberts, D. Kunimoto, J. Manfreda, and R. Long. 1995. Tuberculosis as a primary cause of respiratory failure requiring mechanical ventilation. Am J Respir Crit Care Med 151:867-72.

18. Heffner, J. E., C. Strange, and S. A. Sahn. 1988. The impact of respiratory failure on the diagnosis of tuberculosis. Arch Intern Med 148:1103-8.

19. Ahuja, S. S., S. K. Ahuja, K. R. Phelps, W. Thelmo, and A. R. Hill. 1992. Hemodynamic confirmation of septic shock in disseminated tuberculosis. Crit Care Med 20:901-3.

20. Clark, A. M., W. J. Burman, D. L. Cohn, and P. S. Mehler. 1998. Septic shock from Mycobacterium tuberculosis after therapy for Pneumocystis carinii. Arch Intern Med 158:1033-5.

21. Pene, F., T. Papo, L. Burudy-Gulphe, A. Cariou, J. C. Peiette, and C. Vinsonneau. 2001. Septic shock and thrombotic microangiopathy due to Mycobacterium tuberculosis in a nonimmunocompromised patient. Arch Intern Med 161:1347-8.

22. George, S., L. Papa, L. Sheils, and C. R. Magnussen. 1996. Septic shock due to disseminated tuberculosis. Clin Infect Dis 22:188-9.

23. Crump, J. A., and L. B. Reller. 2003. Two decades of disseminated tuberculosis at a university medical center: the expanding role of mycobacterial blood culture. Clin Infect Dis 37:1037-43.

24. Guttman, P. H. 1930. Addison's desease -- a statistical analysis of 566 cases and a study of the pathology. Arch Pathol 10:742-85.

25. Lam, K. Y., and C. Y. Lo. 2001. A critical examination of adrenal tuberculosis and a 28-year autopsy experience of active tuberculosis. Clin Endocrinol 54:633-9.

26. Barnes, D. J., S. Naraqi, P. Temu, and J. R. Turtle. 1989. Adrenal function in patients with active tuberculosis. Thorax 44:422-4.

27. Annane, D., V. Sebille, G. Troche, J. C. Raphael, P. Gajdos, and E. Bellissant. 2000. A 3-level prognostic classification in septic shock based on cortisol levels and cortisol response to corticotropin. JAMA. 283:1038-45.

28. Annane, D., V. Sebille, C. Charpentier, P. E. Bollaert, B. Francois, J. M. Korach, G. Capellier, Y. Cohen, E. Azoulay, G. Troche, P. Chaumet-Riffaut, and E. Bellissant. 2002. Effect of treatment with low doses of hydrocortisone and fludrocortisone on mortality in patients with septic shock. JAMA 288:862-71.

29. Rogers, R. M., C. Bedrossian, J. J. Coalson, W. W. Cook, J. R. Christiansen, S. Deutsch, R. P. Elkins, B. A. Gray, J. F. Hammarsten, C. Harvey, T. H. Johnson, J. A. Juers, J. A. Mohr, E. R. Rhoades, and G. R. Williams. 1976. The management of massive hemoptysis in a patient with pulmonary tuberculosis. Chest 70:519-26.

30. Cahill, B. C., and D. H. Ingbar. 1994. Massive hemoptysis: assessment and management. Clin Chest Med 15:147-68.

31. Farer, L. S., A. M. Lowell, and M. P. Meador. 1979. Extrapulmonary tuberculosis in the United States. Am J Epidemiol 109:205-17.

32.   Mehta, J. B., A. Dutt, L. Harvill, and K. M. Mathews. 1991. Epidemiology of extrapulmonary tuberculosis. A comparative analysis with pre-AIDS era. Chest 99:1134-8.

33.   Simon, H. B., A. J. Weinstein, M. S. Pasternak, M. N. Swartz, and L. J. Kunz. 1977. Genitourinary tuberculosis: clinical features in a general hospital population. Am J Med 63:410-20.

34.   Goldman, G., and J. L. Bolognia. 1995. Recognizing cutaneous signs of tuberculosis. J Respir Dis 16:646-51.

35.   Fowler, N. O. 1991. Tuberculous pericarditis. JAMA 266:99-103.

36.   Klein, N. C., B. Damsker, and S. Z. Hirschman. 1985. Mycobacterial meningitis: retrospective analysis from 1970 to 1983. Am J Med 79:29-34.

37.   Rieder, H. L., G. D. Kelly, A. B. Bloch, G. M. Cauthen, and D. E. Snider, Jr. 1991. Tuberculosis diagnosed at death in the United States. Chest 100:678-81.

38.   American Thoracic Society. 2003. Treatment of tuberculosis. Am J Respir Crit Care Med 167:603-62.

39.   American Thoracic Society. 2000. Diagnostic standards and classification of tuberculosis in adults and children. Am J Respir Crit Care Med 161:1376-95.

40.   Segura, R. M., C. Pascual, I. Ocana, J. M. Martinez-Vazquez, E. Ribera, I. Ruiz, and M. D. Pelegri. 1989. Adenosine deaminase in body fluids: a useful diagnostic tool in tuberculosis. Clin Biochem 22:141-8.

41.   Aoe, K., A. Hiraki, T. Murakami, R. Eda, T. Maeda, K. Sugi, and H. Takeyama. 2003. Diagnostic significance of interferon-gamma in tuberculous pleural effusions. Chest 123:740-4.

42.   Andersen, P., M. E. Munk, and T. M. Doherty. 2000. Specific imune-based diagnosis of tuberculosis. Lancet 356:1099-104.

43.   Pottumarthy, S., A. J. Morris, A. C. Harrison, and V. C. Wells. 1999. Evaluation of the tuberculin gamma interferon assay: potential to replace the Mantoux skin test. J Clin Microbiol 37:3229-32.

44.   Lalvani, A., A. A. Pathan, H. McShane, R. J. Wilkinson, M. Latif, C. P. Conlon, G. Pasvol, and A. V. Hill. 2001. Rapid detection of Mycobacterium tuberculosis infection by enumeration of antigen-specific T cells. Am J Respir Crit Care Med 163:824-8.

45.   Catanzaro, A., S. Perry, J. E. Clarridge, S. Dunbar, S. Goodnight-White, P. A. LoBue, C. Peter, G. E. Pfyffer, M. F. Sierra, R. Weber, G. Woods, G. Mathews, V. Jonas, K. Smith, and P. Della-Latta. 2000. The role of clinical suspicion in evaluating a new diagnostic test for active tuberculosis: results of a multicenter prospective trial. JAMA 283:639-45.

46.   Broccolo, F., P. Scarpellini, G. Locatelli, A. Zingale, A. M. Brambilla, P. Cichero, L. A. Sechi, A. Lazzarin, P. Lusso, and M. S. Malnati. 2003. Rapid diagnosis of mycobacterial infections and quantitation of Mycobacterium tuberculosis load by two real-time calibrated PCR assays. J Clin Microbiol 41:4565-72.

47.   Hasaneen, N. A., M. E. Zaki, H. M. Shalaby, and A. S. El-Morsi. 2003. Polymerase chain reaction of pleural biopsy is a rapid and sensitive method for the diagnosis of tuberculous pleural effusion. Chest 124:2105-11.

48.   Lima, D. M., J. K. Colares, and B. A. da Fonseca. 2003. Combined use of the polymerase chain reaction and detection of adenosine deaminase activity on pleural fluid improves the rate of diagnosis of pleural tuberculosis. Chest 124:909-14.

49.   Ardito, F., B. Posteraro, M. Sanguinetti, S. Zanetti, and G. Fadda. 2001. Evaluation of BACTEC Mycobacteria Growth Indicator Tube (MGIT 960)

automated system for drug susceptibility testing of Mycobacterium tuberculosis. J Clin Microbiol 39:4440-4.

50. Scarparo, C., P. Ricordi, G. Ruggiero, and P. Piccoli. 2004. Evaluation of the fully automated BACTEC MGIT 960 system for testing susceptibility of Mycobacterium tuberculosis to pyrazinamide, streptomycin, isoniazid, rifampin, and ethambutol and comparison with the radiometric BACTEC 460TB method. J Clin Microbiol 42:1109-14.

51. Small, P. M., and P. I. Fujiwara. 2001. Management of tuberculosis in the United States. New England J Med 345:189-200.

52. Burman, W. J., and B. E. Jones. 2001. Treatement of HIV-related tuberculosis in the era of effective antiretroviral therapy. Am J Respir Crit Care Med 164:7-12.

53. Peloquin, C. A. 1997. Using therapeutic drug monitoring to dose antimycobacterial drugs. Clin Chest Med. 18:79-87.

54. American College of Chest Physicians. 1995. Institutional control measures for tuberculosis in the era of multiple drug resistance. Chest 108:1690-1710.

55. Davis, Y. M., E. McCray, and P. M. Simone. 1997. Hospital infection control practices for tuberculosis. Clin Chest Med. 18:19-33.

56. Bock, N. N., J. E. McGowan, J. Ahn, J. Tapia, and H. M. Blumberg. 1996. Clinical predictors of tuberculosis as a guide for a respiratory isolation policy. Am J Respir Crit Care Med 154:1468-72.

57. Scott, B., M. Schmid, and M. D. Nettleman. 1994. Early identification and isolation of inpatients at high risk for tuberculosis. Arch Intern Med 154:326-30.

58. American Thoracic Society. 2000. Targeted tuberculin testing and treatment of latent tuberculosis infection. Am J Respir Crit Care Med 161:S221-47.

59. von Reyn, C. F., and J. M. Vuola. 2002. New vaccines for the prevention of tuberculosis. Clin Infect Dis 35:465-74.

60. Hoft, D. F., S. Worku, B. Kampmann, C. C. Whalen, J. J. Ellner, C. S. Hirsch, R. B. Brown, R. Larkin, Q. Li, H. Yun, and R. F. Silver. 2002. Investigation of the relationships between immune-mediated inhibition of mycobacterial growth and other potential surrogate markers of protective Mycobacterium tuberculosis immunity. J Infect Dis 186:1448-57.

61. Barnes, P. F., and M. D. Cave. 2003. Molecular epidemiology of tuberculosis. New England J Med 349:1149-56.

62. Delgado, J. C., A. Baena, S. Thim, and A. E. Goldfeld. 2002. Ethnic-specific genetic associations with pulmonary tuberculosis. J Infect Dis 186:1463-8.

63. Lopez-Maderuelo, D., F. Arnalich, R. Serantes, A. Gonzalez, R. Codoceo, R. Madero, J. J. Vazquez, and C. Montiel. 2003. Interferon-gamma and interleukin-10 gene polymorphisms in pulmonary tuberculosis. Am J Respir Crit Care Med 167:970-5.

64. Bellamy, R., M. Beyers, W. J. McAdam, C. Ruwende, R. Gie, P. Samaai, D. Bester, M. Meyer, T. Corrah, M. Collin, D. R. Camidge, D. Wilkinson, E. Hoal-van Helden, H. C. Whittle, W. Amos, P. van Helden, and A. V. S. Hill. 2000. Genetic susceptibility to tuberculosis in Africans: a genome wide scan. Proceed Nat'l Acad Sci. 97:8005-9.

65. Li, C. M., S. J. Campbell, D. S. Kumararatne, R. Bellamy, C. Ruwende, W. J. McAdam, A. V. S. Hill, and D. A. Lammas. 2002. Association of a polymorphism in the P2X7 gene with tuberculosis in a Gambian population. J Infect Dis 186:1458-62.

66. Fairbairn, I. P., C. B. Stober, D. S. Kumararatne, and D. A. Lammas. 2001. ATP-mediated killing of intracellular mycobacteria by macrophages is a P2X(7)-dependent process inducing bacterial death by phagosome-lysosome fusion. J Immunol 167:3300-7.

67.   Saunders, B. M., S. L. Fernando, R. Sluyter, W. J. Britton, and J. S. Wiley. 2003. A loss-of-function polymorphism in the human P2X7 receptor abolishes ATP-mediated killing of mycobacteria. J Immunol 171:5442-6.

68.   Denlinger, L. C., G. Angelini, K. Schell, D. N. Green, A. G. Guadarrama, U. Prabhu, D. B. Coursin, P. J. Bertics, and K. Hogan. 2004. Detection of human P2X7 nucleotide receptor polymorphisms by a novel monocyte pore assay predictive of alterations in LPS-induced cytokine production. Unpublished data.

7

HIV in the ICU

Alan S Karstaedt, L. Rhudo Mathivha, and Christine L. N. Banage
Division of Infectious Diseases and Department of ICU, Chris Hani Baragwanath Hospital and University of the Witwatersrand, Johannesburg, South Africa

INTRODUCTION

At the end of the year 2001, UNAIDS estimated that 25 million of the 36 million people living with HIV infection globally resided in sub-Saharan Africa. The impact of the epidemic in the developing world is devastating. HIV-related mortality is increasing, hospitals are overflowing, the workforce, including healthcare workers, is becoming depleted and the number of orphans is increasing (1). Since the advent of highly active antiretroviral therapy (HAART), HIV infection has been transformed into a treatable infection for people with access to such therapy, with a significant decrease in morbidity and mortality (2-4). Only a tiny fraction of the HIV-infected in the developing world has access to HAART. In the USA and Europe, nearly 75% of HIV-infected people are aware of their serostatus, whereas in the developing world, only 10-20% of HIV-infected people have been tested for HIV (5). Critical care may be required for HIV-associated conditions or opportunistic infections, as well as for unrelated reasons. The challenge for intensive care specialists will be to avoid denying treatment, where it will be beneficial, purely on the basis of HIV infection.

IMPLICATIONS FOR CRITICAL CARE

As more of the HIV-infected population develops complications, the health care system at all levels will be faced by increasing numbers of patients who will present with serious and life threatening conditions. This will also result in an increasing demand for high technology and sophisticated intensive care treatment. Careful and judicious decisions will have to be made about the responsible cost effective utilization of expensive resources. Some of these decisions will raise serious clinical and ethical dilemmas with which each affected institution and country will have to grapple, and which will often be resolved at a political level.

There are a number of serious life-threatening disease complications that necessitate consideration for intensive care intervention. These include pulmonary complications resulting in acute respiratory failure, central nervous system infections such as cryptococcal meningitis, acute systemic bacterial infections that result in severe sepsis and organ compromise, and non-HIV related disease processes such as trauma, post elective surgery ICU care and obstetric complications

ADMISSION CRITERIA AT CHRIS HANI BARAGWANATH HOSPITAL – BROAD GUIDELINES FOR ADMISSION

The following are guidelines that we use to govern admission of patients to a multidisciplinary ICU. The final decision to admit or to refuse admission of a patient to ICU is the prerogative of the ICU consultant on call.

Prospective candidates for ICU admission should have a demonstrable reversible disease process with a reasonable outcome. The general concept that as a consequence of admission to an intensive care unit, positive benefit is expected for the patient is applied.

1. Patients (both paediatric and adult) who have AIDS-defining illnesses are not admitted, as a general rule. The reasons include the poor long-term outlook for HIV-infected people without access to HAART and the poor outcomes of many such patients in an ICU (6).

2. Patients with severe neurological devastation, from whatever cause, are not considered for admission.

3. Patients with end-stage organ or neoplastic diseases are not considered good ICU admission candidates. The exceptions are patients with these conditions who are postoperative for palliative procedures requiring less than 48hrs post-operative observation in ICU.

These criteria, other than criterion 1, are applied to all patients who are being considered for ICU admission and do not just single out HIV/AIDS patients. It has been a difficult necessity to have these stringent criteria due to the small size of the unit i.e 28 beds for a 3000-bed tertiary referral hospital. This scenario is identical in most large hospitals in Southern Africa. Unless specific resources are earmarked for the care of the HIV-infected, including the provision of antiretroviral drugs, patients with AIDS-defining illnesses will continue to be excluded. These criteria would exclude more than half of the patients admitted to ICU in the USA.

We do not demand HIV testing of all possible admissions. Surgical patients, consisting of both elective and trauma cases, are not routinely tested for HIV prior to admission. Similarly, obstetrics and gynecology patients tend to be admitted with non-HIV-related conditions, commonly eclampsia, HELLP syndrome, post-partum hemorrhage and incomplete abortions.

OUTCOME

In the early 1980's initial outcome data on HIV positive patients admitted to our ICU showed mortality rates of between 86-100%. This decreased to 45-62% in the late 1980's (unpublished data). In Zambia the outcome was similar in both HIV-seropositive and seronegative patients (6). The only subgroup with a worse prognosis was the group with HIV-related pneumonia. In the developed world, hospital or ICU mortality rates for patients with HIV infection or AIDS have been 28-51% (7).

SPECIFIC CONDITIONS

Surgical patients
In a Durban surgical intensive care unit, a prospective double blind study of all admissions over 6 months investigated the influence of HIV status on the ICU admission and outcome (8). No patients had AIDS. The HIV-infected compared to the HIV-non-infected were similar by gender but were younger. They had significantly more organ failure (71% compared to 49% of patients) and a higher incidence of septic shock (38% compared to 15%). There were no differences in the duration of stay in the ICU or in ICU or in-hospital mortality.

This was a controversial article in that informed consent was not obtained from the patients for participation in the study or for testing for HIV antibodies. A debate ranged around both the question of whether such a study should be performed at all despite the importance of the research question (9-11) and also whether a reputable journal should publish such a study (12-14).

Gynecological conditions
We have noted that those with underlying HIV infection tend to develop more nosocomial sepsis than those who are HIV negative, thereby prolonging their ICU stay. In Zimbabwe, there was no significant difference in outcome by HIV status (15).

Burns
Little is found in the literature concerning burn injuries in patients with HIV infection. A study from South Africa compared 33 HIV-infected patients with moderate to severe burns affecting a mean total body surface area of 24.6% with matched HIV negative controls. Matching was for age, gender, degree of burns and inhalation injury. There was no significant difference in mortality or treatment parameters. The two patients with AIDS (and tuberculosis) died. The authors concluded that an HIV-infected patient with burns and without AIDS could be treated similarly to HIV negative patients with similar outcomes expected (16).

Guillain-Barre Syndrome
The demyelinating neuropathies are the most common neurologic complications occurring during the clinically latent phase of HIV infection. These are similar to the subacute Guillain-Barre syndrome or chronic inflammatory demyelinating polyneuropathy seen in HIV negative patients, with the exception that the cerebrospinal fluid often reveals an unexpected pleocytosis (17,18). An autoimmune pathophysiology is suggested by a perivascular infiltrate on peripheral nerve biopsy. Intensive care admission is usually based on respiratory compromise. Treatment is with intravenous immunoglobulins or plasma exchange. The prognosis of HIV infected patients may not be as good as HIV negative patients (19,20), although a small study from Johannesburg suggests an equivalent or even better outcome in the HIV-infected (21).

Highly Active Antiretroviral Therapy (HAART)
Although most HIV-infected people in the developing world cannot afford HAART at present, there is a growing hope and expectation that such treatment will become affordable. Countries like Brazil and Botswana lead the way in promoting widespread access for their populations. The

World Health Organisation (WHO) has declared universal access to therapy to be a goal.

Patients with severe toxicity, such as pancreatitis and nucleoside-associated lactic acidosis, from antiretroviral drugs may need ICU admission. Severe lactic acidosis, which was first described in 1991 in association with the use of didanosine (22), occurs in 2-4 people per 1000 patient-years (23). The mechanism of the nucleoside reverse transcriptase inhibitor (NRTI) – mediated toxicity is mitochondrial toxicity that is ascribed to the polymerase gamma hypothesis (24). DNA polymerase gamma is the only polymerase involved in mitochondrial DNA replication. Nucleoside analogues inhibit mitochondrial DNA polymerase gamma. This leads to a reduction in mitochondrial DNA and alteration in the synthesis of mitochondrial proteins. The loss of mitochondrial oxidative function leads to increased reliance on anaerobic metabolism and the result is an accumulation of lactate and thus of acid. Although all NRTI's have been implicated in cases of lactic acidosis, most have been associated with stavudine use. The syndrome occurs 1-20 months after starting NRTI therapy. Risk factors include female gender, high body mass index and concurrent liver disease. It can occur at any stage of HIV infection. The symptoms are abdominal including abdominal pain and distention, nausea, vomiting and anorexia, and fatigue and weight loss. There may be biochemical hepatocellular dysfunction and a lactate level >5 mmol/L. The mortality rate is very high, ranging from 37%-57% of patients (25-26). A serum lactate > 10 mmol/L correlates with mortality. HAART should be immediately stopped in the presence of symptoms and a lactate > 5mmol/L. Administration of essential cofactors such as thiamine, riboflavin, L-carnitine vitamin C and antioxidants have been used as therapy for congenital mitochondrial diseases. A few case reports have suggested benefit in lactic acidosis (27).

Antiretroviral drugs inhibit cytochrome P450 3A4 and thereby may interfere with the metabolism of many drugs used in the ICU. The risks and benefits of stopping HAART in the ICU setting are largely unmeasured (28). Only one drug has a parenteral formulation and enteral administration may lead to poor absorption of drugs.

CD4 T-Lymphocytes

Some clinicians use the CD4 T-lymphocyte count as a surrogate marker for HIV infection, especially when informed consent for HIV testing is not possible, or as a factor in deciding on ICU admission and management. In a study from Oakland, California, the CD4-cell counts of

102 consecutive patients (72 medical and 30 surgical) admitted to ICU were studied (29). Only 3 patients tested positive for HIV antibodies. Yet 41% of patients had CD4 counts of <400 cells / mm^3 and 29% had CD4 counts of <300 cells / mm^3. CD4 counts were linearly related to total lymphocyte counts, which could therefore be predictive of CD4 count in resource poor settings. There was no correlation between CD4-cell count and APACHE II score, predicted mortality rate or survival. Thus, acute illness alone in HIV non-infected patients can be associated with marked decreases in CD4 counts. No conclusions regarding HIV serostatus or survival can be based on single measurements in acutely ill patients.

Advance directives
Patient wishes for life-sustaining treatment also need to be taken into consideration. A person's right to autonomy includes the right of a mentally competent and informed person to refuse intensive care with ventilation. This advance directive can be made in the form of a living will or by designating another person with durable power of attorney to make such a decision should the need arise. This should also be discussed with the person's primary care giver. Such advance directives are of less practical utility in resource-poor countries, with limited ICU facilities which are often denied to patients with AIDS, compared to the developed world.

OCCUPATIONAL EXPOSURE TO HIV INFECTION

The entry of HIV positive patients into the invasive, close contact ICU environment can lead to an increase in the occupational risk of exposure to the disease as health care workers come into contact with potentially infective blood and other body fluids.

The occupational risk of HIV infection for health care workers is based on three factors (30). The first is the risk of needlestick injury or mucous membrane exposure per year. The second is the risk of HIV transmission from a percutaneous exposure to HIV-infected blood, which is estimated to be 0.3% [CI is 0.2%-0. 5%], or from a mucous membrane exposure, which is 0.09% [95% CI = 0.006%-0.5%]. In a retrospective case-control study of health care workers with percutaneous exposure to HIV, the risk for HIV infection was increased with exposure to a larger quantity of blood as indicated by visibly bloody sharp devices, a needle having been placed in a blood vessel, or a deep injury. The risk was also increased for blood from patients with preterminal AIDS (death within 2 months). The use of zidovudine as post-exposure prophylaxis reduced the risk of HIV infection by 81% [95%CI=43%-94%] (31). The limitations of this study include the small number of cases studied and the use of cases and controls from different cohorts. The third factor is the prevalence of HIV infection in the population served. This last factor is the cause of increased risk in the many developing countries with high rates of HIV infection. Among interns at two academic hospitals in Johannesburg, South Africa, 83% recalled a percutaneous injury during medical school or internship, and 43% recollected a percutaneous injury to HIV-infected blood. Of the interns, 54% had experienced a mucocutaneous exposure to HIV-seropositive blood. In total, 70% had a blood exposure to an HIV-positive source by the end of the internship year (32).

In order to safeguard healthcare workers, strict application of universal precautions to all patients is required. Other measures, which can be applied in an ICU setting, include moving to a needle-less unit as far as possible. Counseling and post-exposure prophylaxis should be readily available

The risk of transmission from HIV-infected health care personnel to patients is extremely low. There have been only 2 published reports of such transmission, one of a dentist in the United States in 1990 (33) and one of an orthopaedic surgeon in France in 1997 (34). Infected health care workers should ideally be under medical and occupational health supervision.

CONCLUSION

In severely resource-limited settings, a number of questions will continue to vex ICU clinicians:
1. Should a patient's HIV seropositivity disqualify the person from admission into an ICU? Individual hospitals and regional groups need to identify poor prognostic features and develop their own triage guidelines.
2. Most developing countries are not able to provide the patient with maintenance HAART once discharged from the ICU. Should this situation change, will providing expensive high-technology treatment to previously unrecognized as well as known AIDS patients become standard of care in developing countries?
3. How can one develop practices that do not discriminate against the HIV-infected while not squeezing out the HIV-non-infected?

4. What should be the relation and balance between a country's efforts at preventing HIV infection and providing adequate care for those who have been infected?

REFERENCES

1. Piot P, Bartos M, Ghys PD, Walker N, Schwartlander B. The global impact of HIV/AIDS. Nature 2001; 410: 968-973.
2. Moore RD, Chaisson RE. Natural history of HIV infection in the era of combination antiretroviral therapy. AIDS 1999; 13: 1933-1942.
3. Pallela FJ, Delaney KM, Moorman AC, et al. Declining morbidity and mortality among patients with advanced human immunodeficiency virus infection. N Engl J Med 1998; 338: 853-860.
4. Mocroft A, Vella S, Benfield TL, et al. Changing patterns of mortality across Europe in patients with human immunodeficiency virus infection. Lancet 1998; 352: 1725-1730.
5. United Nations AIDS Programme. HIV/AIDS: the global epidemic. Geneva: United Nations AIDS Programme, 1998.
6. Nickas G, Wachter RM. Outcomes of intensive care for patients with human immunodeficiency virus infection. Arch Intern Med 2000; 160: 541-547.
7. Watters DAK, Wilson IH. HIV infection, AIDS and ICU in Central Africa. Anaesthesia 1988; 43: 988-989.
8. Bhagwanjee S, Muckart DJJ, Jeena PM, Moodley P. Does HIV status influence the outcome of patients admitted to a surgical intensive care unit? A prospective double blind study. BMJ 1997; 314: 1077-1081.
9. Kale R. Commentary: failing to seek patients' consent to research is always wrong. BMJ 1997; 314:1081-1082.
10. Bhagwanjee S, Muckart DJJ, Jeena PM, Moodley P. Commentary: why we did not seek informed consent before testing patients for HIV. BMJ 1997; 314:1082-1083.
11. Seedat YK. Commentary: no simple and absolute ethical rule exists for every conceivable situation. BMJ 1997; 314: 1083-1084.
12. Doyal L. Journals should not publish research to which patients have not given fully informed consent-with three exceptions. BMJ 1997; 314: 1107-1111.
13. Tobias JS. BMJ's present policy (sometimes approving research in which patients have not given fully informed consent) is wholly correct. BMJ 1997; 314: 1111-1114.
14. Anonymous. All treatment and trials must have informed consent. BMJ 1997; 314: 1134-1135.
15. McKenzie AG. Outcome of patients with human immunodeficiency virus (HIV) infection admitted to intensive care: a preliminary study. Cent Afr J Med 1991; 37: 436-437.
16. Edge JM, Van der Merwe AE, Pieper CH, Bouic P. Clinical outcome of HIV positive patients with moderate to severe burns. Burns 2001; 27: 111-114.
17. Cornblath D. Treatment of the neuromuscular complications of human immunodeficiency virus infection. Ann Neurol 1988; 23 (Suppl): S88-S91.
18. Cornblath D, McArthur J. Predominantly sensory neuropathy in patients with AIDS and AIDS-related complex. Neurology 1988; 38: 794-796.

19.   Cornblath D, Chaudhry V, Griffin J. Treatment of chronic inflammatory demyelinating polyneuropathy with intravenous immunoglobulin. Ann Neurol 1991; 30: 104-106.

20.   Simpson DM, Olney RK. Peripheral neuropathies associated with human immunodeficiency virus infection. Neurol Clin 1992; 10: 685-711.

21.   Schleicher GK, Richards GA, Black A. The effect of HIV on the outcome of patients admitted to the ICU with Guillain-Barre syndrome. Abstract 58; Combined Congress of the South African Thoracic Society and Critical Care Society of South Africa, Sun City 2002.

22.   Lai KK, Gang DL, Zawacki JK, Cooley TP. Fulminant hepatic failure associated with 2,3,-didexyinosine (DDI). Ann Intern Med 1991; 115: 283-284.

23.   John M, Moore CB, James IR, et al. Chronic hyperlactatemia in HIV-infected patients taking antiretroviral therapy. AIDS 2001; 15: 717-723.

24.   Brinkman K, ter Hofstede HJ, Burger DM, et al. Adverse effects of reverse transcriptase inhibitors: mitochondrial toxicity as common pathway. AIDS 1998; 12: 1735-1744.

25.   Miller KD, Cameron M, Wood LV, Dalakas MC, Kovacs JA. Lactic acidosis and hepatic steatosis associated with use of stavudine: report of four cases. Ann Intern Med 2000; 133: 192-196.

26.   Falco V, Rodriguez D, Ribera E, et al. Severe nucleoside-associated lactic acidosis in human immunodeficiency virus-infected patients: report of 12 cases and review of the literature. Clin Infect Dis 2002; 34: 838-846.

27.   Brinkman K, Vrouenraets S, Kauffmann R, Weigel HM, Frissen PHJ. Treatment of nucleoside reverse transcriptase inhibitor-induced lactic acidosis. AIDS 2001; 14: 2801-2802.

28.   Soni N, Pozniak A. Continuing HIV therapy in the ICU. Crit Care 2001; 5: 247-248.

29.   Feeney C, Bryzman PA, Kong L, et al. T-lymphocyte subsets in acute illness. Crit Care Med 1995; 23: 1680-1685.

30.   Centers for Disease control and Prevention. Updated U. S. Public Health Service guidelines for the management of occupational exposures to HBV, HCV, and HIV and recommendations for post-exposure prophylaxis. MMWR 2001; 50: (No RR-11): 1-52.

31.   Cardo DM, Culver DH, Ciesielski CA, et al. A case-control study of HIV seroconversion in health care workers after percutaneous exposure. N Engl J Med 1997; 337: 1485-1490.

32.   Karstaedt AS, Pantanowitz L. Occupational exposure of interns to blood in an area of high HIV seroprevalence. S Afr Med J 2001; 91: 57-61.

33.   Ciesielski CA, Marianos D, Ou CY, et al. Transmission of HIV in a dental practice. Ann Intern Med 1992; 116: 798-805.

34.   Transmission of HIV from an infected surgeon to a patient in France. Communicable Disease Report Weekly 1997 7: 17.

8

HIV and Associated Infections in the Intensive Care Unit: Perspectives from North America

Scott E. Evans and Andrew H. Limper
Division of Pulmonary, Critical Care Medicine and Internal Medicine, Mayo Clinic and Foundation, Rochester, Minnesota 55905

1. INTRODUCTION

Asia and Africa have experienced a degree of devastation from human immunodeficiency virus/acquired immunodeficiency syndrome (HIV/AIDS) not encountered within the United States. Still, in the setting of an estimated 40 million people living with AIDS worldwide, the United States has approximately 900,000 cases since 1981. At the end of 2001, a reported 43,158 people were living with AIDS in the United States (1,2).

The cumulative medical burden of HIV is difficult to assess, in part due to incomplete surveillance for HIV seropositive patients without AIDS. However, the total utilization of inpatient medical services for HIV-related disease appears to have decreased since the broad acceptance of multi-drug highly active antiretroviral therapy (HAART) in 1996-7 (3-5). Commensurately lessened disease complications and AIDS-related deaths have been documented for all HIV risk groups in the United States since that time (1,6). Similar trends in admissions, morbidity, and mortality have also been observed for critical care services provided to HIV-infected patients. Currently, an estimated 4 to 12% of hospitalizations for HIV-infected patients require ICU admission (7-9).

The utilization of intensive care unit (ICU) services for HIV-infected patients has varied widely since the beginning of the AIDS epidemic. The early 1980's witnessed unacceptably poor ICU outcomes with in-hospital mortality reported as high as 69% for HIV-infected patients requiring ICU-admission (10), with estimated costs of >$305,000 per life year gained by ICU care (11). Ethical debates ensued about the futility and distributive justice of ICU management for HIV-infected patients, presumably impacting attitudes about the critical care of AIDS patients in manners that may still persist today (7,12).

Wachter and colleagues have since defined five "eras" in the ICU care of HIV-infected patients. From retrospective reviews of San Francisco General Hospital ICU admissions, these authors have noted improving survival to hospital discharge from only 31% in Era I (1981-1985)(10) to 71% in Era V (1996-1999), with 25% of the latter cohort receiving HAART upon ICU admission (13). This improvement appears also to be associated with a modest but significant increase in median post-hospitalization survival rates (13,14).

The reasons for improved hospital mortality rates are complex and the relevant literature is limited by its largely retrospective nature. Curtis and colleagues have also documented dramatic variability in ICU utilization for HIV-infected patients across the United States (15,16). Such studies have emphasized the marked impacts of geographic location, hospital ownership/profit-status, and hospital HIV experience on ICU admission decisions. In a related review, Rosen convincingly argues that the literature's outcomes data are skewed by local policies regarding ICU admission of HIV-infected patients (12). It has further been suggested that outcome variability may also represent shifts in clinicians' attitudes reflecting court decisions regarding the withholding of care for patients with HIV infection.

After considering for these confounding factors, continually evolving sources of patient and practice heterogeneity still render even prospective investigations unlikely to provide comprehensive understanding of critical care among HIV infected patients. These presently include variations in immunological responses to HAART (17), changing demographics of HIV infection and access to health care (13), and modifications in HIV surveillance (1).

2. REASONS FOR ICU ADMISSION IN HIV-INFECTED PATIENTS

Just as with patients without HIV, the need for ICU admission arises from innumerable conditions among patients infected with HIV/AIDS. Hauser characterizes the most common precipitants of ICU admissions among HIV-infected patients as (1) respiratory insufficiency, (2) cerebral dysfunction, and (3) hypotension (18). In addition, just as in patients without AIDS, the underlying causes of clinical presentations may be entirely unrelated to the patient's HIV status. However, even common syndromes caused by non-opportunistic infections must be addressed in a manner tailored to the special needs of the HIV-infected patient.

2.1. Non-AIDS-Related Diagnoses

HIV-infected patients require ICU admission more often than the general population, yet most ICU admissions are not attributable to a specific AIDS-related diagnosis. That is, while the most common specific diagnoses prompting ICU admission are directly related to HIV, the majority of ICU admissions actually arise from a broad array of different non-AIDS-related causes (7,9,13). Nonetheless, whether admitted for routine post-operative management or for an acute coronary syndrome, the care of HIV-infected patients in the ICU demands unique attention to certain principles.

In addition to an awareness of a patient's HIV status, all direct care providers must possess a fundamental understanding of the mechanisms and rates of HIV transmissibility. This includes an appreciation for the actual risk that they encounter in caring for HIV-infected patients (which may be overestimated among providers inexperienced in caring for this population). This knowledge not only helps protect the ICU care team from exposure through enhanced adherence to universal precautions, but also avoids undue isolation and unwarranted withholding of services which providers incorrectly perceive as dangerous.

The ICU care team must also be cognizant of potential untoward drug effects observed in HIV-infected patients. Complex medical regimens place this patient population at elevated risk for drug-drug interactions, including interactions between multiple anti-HIV therapies or between anti-retroviral agents and procedure-related medications. Of particular note is the metabolic interdependence of protease inhibitors with the cytochrome P450 system, which may be altered by numerous drugs given in the operating room or ICU. Similarly, drugs administered in the outpatient setting (e.g., probenecid) may sufficiently alter antiretroviral pharmacokinetics to produce unanticipated drug interactions in the ICU. Further, the toxicities associated with HAART enhance the likelihood that overlapping or cumulative side effect profiles will render patients intolerant of commonly used ICU medications (19).

Additionally, ICU care providers must be equipped to determine whether critically ill patients should receive antiretroviral therapies, an issue not without controversy. For some patients, the decision is foregone as all currently approved antiretroviral agents are delivered as oral preparations. For other patients, the intensivist is required to weigh potential risks and benefits. Some suggest that attendant drug toxicities warrant the discontinuation of HAART while in the ICU, but reports of cultivated resistance and insurmountable viral load rebounds following intermittent antiretroviral use temper this position (20). For patients naïve to therapy, reports exist of a so-called "immunological reconstitution syndrome" with paradoxically heightened susceptibility to infections including tuberculosis and *Pneumocystis* pneumonia after initiation of HAART (21-23). However, short-term complications associated with initiation of HAART have not correlated with worse long term outcomes, with some authors describing improved outcomes from *Pneumocystis* pneumonia after initiating HAART (24).

Aside from these issues, the general management of HIV-infected patients in the ICU is similar to that of seronegative patients. However, because as many as 27.6% of initial HIV diagnoses may be first made in the ICU (24), even if the precipitating event for admission is not AIDS-related, the ICU care teams must remain prepared to address these issues, regardless of the subspecialization of their individual units.

2.2. AIDS-Related Diagnoses

Complications of HIV/AIDS result in ICU admission through diverse mechanisms, resulting from both direct viral effects and increased susceptibility to other conditions. These are reviewed below, with emphasis on the most frequent causes of AIDS-related morbidity and mortality.

2.2.1. HIV-Associated Organ Damage. HIV directly damages many organ systems, either through direct cellular toxicity or secondary to associated host inflammatory responses. This may be manifest as HIV-induced myocarditis/cardiomyopathy, myelopathy, enteropathy, encephalopathy/dementia, interstitial pneumonia, emphysema, dermatitis/pruritus, as well as affective and psychotic disorders (25-35). In addition, conditions such as HIV-induced anemia and thrombocytopenia appear to be mediated by humoral immunity, with target damage arising from antigen-antibody complexes rather than from cytokine-induced inflammation (36).

While these and other examples of HIV-induced organ damage are well characterized in the literature, their identification remains difficult in the ICU. Each may be sufficiently severe to require ICU management, yet none may be diagnosed except by exclusion. As a group, these conditions have convincing (and often multiple) clinical imitators, typically related to opportunistic infections or toxic medication effects. Furthermore, clinical signs such as pulmonary infiltrates may be caused by several of these entities, by opportunistic infection, by drug toxicities, or by any number of non-HIV-related conditions.

Once established, some HIV-induced organ damage, such as cardiomyopathy and dementia, are not significantly reversible, though effective antiretroviral therapy may halt progression. In contrast, others conditions (e.g., enteropathy, pruritus) potentially resolve with therapy. Thus, *bona fide* HIV-induced organ damage may be a realistic ICU indication for institution of HAART in the absence of infectious contraindications. Corticosteroids play an unclear role in the management of HIV-induced organ damage, though they appear to be beneficial in the management of immune mediated anemia and thrombocytopenia, in concert with intravenous immunoglobulin and splenectomy.

2.2.2. HIV-Related Drug Toxicities. Just as HIV-induced organ damage can result in ICU admissions, so too can the untoward effects of antiretroviral therapies. For example, 2',3'-dideoxinosine (ddI) may result in peripheral neuropathy at therapeutic doses. Similarly, stavudine (d4T) therapy may result in mitochondrial toxicity with lactic acidosis. To further complicate this problem, the use of either of these drugs along with parenteral pentamidine may result in severe pancreatitis (19). Therefore, the intensivist must diligently review the medication lists of HIV-infected patients when attempting to explain unexpected clinical syndromes. And, perhaps more importantly, he or she must insure sufficient expertise or access to reference sources to avoid unintentional complications. With increasing reliance on multi-agent HAART, this consideration is expected to hold increasing importance in the foreseeable future.

2.2.3. AIDS Associated Malignancies. The impaired immune surveillance for altered host cells characteristic of AIDS predisposes to an increased incidence of opportunistic malignancies in this patient population. Non-Hodgkin lymphoma accounts for the greatest number of AIDS-associated malignancy-related deaths, and thus commonly results in ICU admissions (37,38). While the advent of HAART is postulated to modify the epidemiology of this disease, the intensivist will continue to manage acute complications of such malignancies and to facilitate hematologic evaluation prior to administration of cytotoxic chemotherapy. Though a less common cause of ICU admissions, extracutaneous Kaposi sarcoma (KS) presents unique challenges to the intensivist. First, he or she must consider this in the differential when evaluating unexplained lesions such as pulmonary nodules or gastrointestinal masses. Next, critical decisions must be made regarding therapy. While up to 86% of KS lesions may resolve with HAART, the intensivist must consider whether the lesions are relevant to the acute crisis. In addition, it should be decided whether the more rapid response of cytotoxic chemotherapeutics, such as paclitaxel or doxorubicin is required, and whether or not the patient can tolerate the toxicities of either HAART or chemotherapy (38).

2.2.4. Opportunistic Infections in the ICU. More common than opportunistic malignancies, however, are opportunistic infections among AIDS patients. While severity of illness places all ICU patients at increased risk of infection, almost no other group is comparably susceptible to infection by otherwise non-pathologic opportunistic organisms. The spectrum of opportunistic organisms to which AIDS patients are susceptible, when considered with their concomitant vulnerability to community-acquired and nosocomial pathogens, demands that antimicrobial therapy be based on specific organism identification and culture.

Because of the rapid therapeutic response often essential in the ICU, it is reasonable to approach AIDS-related infectious crises with empiric antimicrobial agents targeted at the most likely responsible organism while awaiting laboratory and culture results. The clinician can often be directed to the correct initial regimen by patient characteristics. Although not sufficient to consistently allow empiric therapy, there remains a surprisingly tight correlation between total $CD4^+$-cell counts and infecting opportunistic organisms (39,40). Knowledge of geographic distribution of infectious agents can further assist an intensivist in estimating the likely causes of acute infectious crises. Understanding of other patient defense impairments, such as impaired gag, can further assist with empiric management. Definitive care of some of the most often encountered organisms is discussed below.

2.2.4.1. Pneumocystis Pneumonia. Despite improved awareness of prophylaxis guidelines, *Pneumocystis* pneumonia remains both the most frequent cause of HIV-related respiratory insufficiency and the single most common ICU admitting diagnosis among AIDS patients. Conclusive figures are lacking due to literature heterogeneity, but HAART-era studies indicate that *Pneumocystis* pneumonia accounts for between 10.7 and 24.0% of ICU admissions among patients with AIDS (7,13,41). As many as 24% of hospitalized AIDS patients may require ICU admission for *Pneumocystis* pneumonia (7). These figures reflect a reduction in the percent of admissions, and fortunately, outcomes have improved since the dismal survival rates first observed in the 1980s.

Patients receiving HAART therapy appear less likely to suffer from *Pneumocystis* pneumonia (33), and more likely to survive ICU admissions for *P. carinii*-induced respiratory failure (25% mortality vs. 63% mortality for non-HAART treated patients in one recent study) (24). However, the most dramatic improvements in *Pneumocystis* pneumonia survival relate not to HAART, but to the use of adjunctive corticosteroid therapy since the mid-1980s. Numerous investigators have demonstrated improvements in markers of respiratory failure, oxygenation, and mortality when adjunctive corticosteroids are added to anti-*Pneumocystis* agents (42,43). This phenomenon presumably reflects a reduction in lung damage resulting from exuberant host defenses against *Pneumocystis*, a hypothesis supported by observations that neutrophilic invasion strongly predicts mortality from this infection, to a far greater degree than does organism burden (33,44,45). Current studies suggest nearly universal use of adjunctive steroids in the ICU (16,24). This is appropriate for moderate to severe disease in the absence of contraindications, such as uncontrolled cytomegalovirus (CMV) co-infection or prior severe intolerance to steroid administration.

Conversely, the benefits of corticosteroids are highly questionable in mild disease ($P_aO_2 > 70$ mmHg), and some have indicated that their benefit is lost if not initiated within 72 hours of respiratory failure onset (42).

The need for mechanical ventilation consistently presages worsened outcomes for AIDS patients with *P. carinii* pneumonia, regardless of other therapies, with hospital mortality in this group reported between 40 and 90% (8,14,46,47). Specific reasons for this high mortality are unclear, but this population is recognized at elevated risk of spontaneous pneumothorax as well as acute respiratory distress syndrome and ventilator-induced lung injury. No ventilator strategies specific to *P. carinii* pneumonia have yet been identified which improve outcome. Preliminary investigations into non-invasive positive pressure ventilation (NPPV) have suggested decreased need for intubations and improved survival when continuous positive airway pressure was delivered by facemask (48). The reasons for this difference have not yet been established, and recommendations for NPPV will require further investigation.

Anti-*Pneumocystis* therapies have changed little in recent years. Studies continue to indicate that most patients receive trimethoprim-sulfamethoxazole (TMP-SMX) as first line therapy. Clinical response of *Pneumocystis* to TMP-SMX and other agents is slow, and therapy should be continuing for three weeks in most cases. Although laboratory data suggest continued organism susceptibility in most patients, up to half of ICU patients on TMP-SMX will be changed to alternate therapies, perhaps related to this slow effect of therapy (24). Parenteral pentamidine should be considered for patients intolerant of TMP-SMX. Patients intolerant to both, or who fail to improve after one week of therapy, may benefit from trimetrexate-leucovorin combination therapy. Common dosing regimens for these medications are provided in Table 1. Atovaquone is generally reserved for patients with mild to moderate disease.

2.2.4.2. Mycobacterial Infections. The extreme susceptibility of HIV-infected patients to *Mycobacterium tuberculosis* infection poses an important challenge to the ICU team (49,50). AIDS patients in the ICU with respiratory insufficiency syndromes consistent with tuberculosis should routinely remain in respiratory isolation pending three sputum samples demonstrated free of acid-fast bacilli. Nucleic acid amplification for *M. tuberculosis* is now rapidly available in many centers and dramatically increases the sensitivity of sputum or respiratory secretion evaluation. Exclusion of *M. tuberculosis* may prove pragmatically difficult, owing to the numerous potential presentations of pulmonary tuberculosis. However, the propensity of ICU procedures to result in aerosolization of respiratory secretions (e.g., bronchoscopy or intubation) places both ICU staff and other critically ill patients at risk for exposure to considerable *M. tuberculosis* inocula.

M. tuberculosis constitutes a very large percentage of acid-fast bacilli identified in respiratory secretions of AIDS patients in the ICU, and causes disease while CD4$^+$-cell counts remain in the normal range. When detected, *M. tuberculosis* always requires multi-drug therapy. Conversely, culture-proven *M. avium intracellulare* typically does not cause infections in HIV patients except at very low levels of CD4$^+$-cells, even if identifiable by smear. Without compelling evidence to the contrary, therapy for *M. avium intracellulare* may be withheld unless CD4$^+$-cell counts are less than 100/mm^3. The relevance of less frequently cultured mycobacteria (*M. kansasii, M. gordonae, M. bovis*) remains unclear. Though these organisms cause disease less often, antimycobacterial therapy decisions must consider the individual patient. It is our opinion, that both *M. kansasii* and *M. bovis* should be considered potentially aggressive and may well merit treatment.

2.2.4.3. Other Pneumonias. The broad range of potential pathogens in AIDS-related pneumonia demands prompt culture of either acceptably collected sputum or bronchoalveolar lavage fluid. The heterogeneous syndromes produced by bacterial pneumonias during AIDS threaten not only respiratory function, but also result in a staggeringly high incidence of sepsis (51). Initial empiric antibiotics must consider geographic and patient characteristics and should include coverage for *Streptococcus pneumoniae*, *Haemophilus influenzae*, and atypical bacteria such as *Legionella pneumophila*, or *Chlamydia pneumoniae*, and *Mycoplasma pneumoniae*. In the ICU-setting, many such patients have also received recent anti-microbial therapy or recent institutional care. Therefore, the possibilities of drug resistant *Streptococcus pneumoniae* and *Pseudomonas aeruginosa* should be considered when implementing empiric antibiotic coverage.

Even broad-spectrum empiric antibiotic regimens yield unacceptably high failure rates when used as definitive therapy for several reasons. First, the offending pathogens often include not only bacteria, but also may involve viruses, fungi, and potentially even protozoa or helminths. Secondly, even if infected by common bacterial organisms, prior infections and therapies often select multiply resistant bacteria. Third, low to intermediate organism antibiotic susceptibilities, which might be sufficiently bacteriostatic to allow immunocompetent patients to suppress infections, may be inadequately bacteriocidal for patients with the immune derangements of AIDS.

Along with retinitis and enteritis, CMV pneumonia represents a significant source of morbidity among profoundly immunosuppressed AIDS patients. Both ganciclovir and foscarnet are acceptable choices for proven or suspected CMV pneumonia. Notably, the ubiquitous presence of CMV in respiratory secretions in patients with <50-100 $CD4^+$ cells/mm^3 often results in positive viral cultures without active pulmonary disease. Consequently, transbronchial or open lung biopsy are required for definitive diagnosis. During the winter months, influenza respiratory infection should also be considered. If clinically appropriate, amantadine, rimantadine, zanamivir, or oseltamivir may be considered for treatment of influenza in these immune suppressed patients. Acyclovir may be used if varicella pneumonia is suspected.

Despite geographic variations, when fungal pneumonia is considered and there are no contraindications, amphotericin B is a reasonable initial empiric therapy. Amphotericin B provides adequate initial coverage for *Histoplasma capsulatum*, *Aspergillus fumigatus*, *Blastomyces dermatitidis*, and *Coccidiodes immitis*, with subsequent azole therapy directed by eventual definitive organism identification. While the addition of flucytosine to amphotericin B is preferable for initial *Cryptococcus neoformans* management, the single agent still typically provides sufficient coverage during the diagnostic evaluation.

2.2.4.4. Central Nervous System (CNS) Infections. In addition to HIV encephalopathy, AIDS patients may be admitted to the ICU with neurological changes resulting from opportunistic or community-acquired CNS infections. The most typical changes are altered mental status, new onset seizures, focal motor or sensory deficit and meningitis.

Diffuse encephalopathies may be observed from progressive multifocal leukoencephalopathy (PML), CMV encephalitis or herpes simplex (HSV) encephalitis. No therapies besides HAART have been shown to improve PML outcomes, though ganciclovir or foscarnet may be of benefit in CMV and Acyclovir may improve HSV outcomes. After radiographic exclusion of mass lesion or hydrocephalus, cerebrospinal fluid (CSF) collection for suspected AIDS-related meningitis is indicated by relatively effective therapies and reasonably high diagnostic yield of the procedure. Among the typically identified organisms are *Cryptococcus*, and *M. tuberculosis.* CSF analysis may also indicate the presence of lymphoma, another common cause of AIDS-related meningitis.

CNS mass lesions may represent either malignancy or infection. Because of its frequency, *Toxoplasma gondii* should be considered in patients with <100 $CD4^{+}$-cells/mm^3 and new intraparenchymal lesions, even in the absence of a classic ring-enhancing lesion. Alternate considerations include bacterial abscesses, lymphoma, and tuberculosis. Similarities of presentation may eventually warrant brain biopsy, though this decision is best made in consultation with infectious disease and neurosurgical subspecialists.

3. PREDICTORS OF ICU OUTCOMES AMONG HIV-INFECTED PATIENTS

Numerous efforts have been made to establish prognostic markers for critically ill AIDS patients. Table 2 reviews some of these studies, with comment on patient population and study results.

To date, no measures are known to effectively predict ICU mortality among HIV-infected patients (15). The hypothesis follows that severity of illness, rather than HIV-related factors, primarily determines short-term outcomes. This is cautiously supported by observations that acute physiology scores (e.g., APACHE II or SAPS I and II) tend to outperform demographic or laboratory data in predicting outcome (51-54). Still, in many reports APACHE II scores underestimate ICU mortality in HIV disease. Further, trends suggest that non-AIDS diagnoses portend better outcomes than do admissions for *P. carinii*-related respiratory insufficiency, even when correcting for severity of illness (13). Thus, no predictors of HIV-related ICU admission outcome, have been conclusively identified.

Even in the presence of a validated scoring system for HIV outcomes, the principals of care remain the same. Team members must control nosocomial infection risk, minimize untoward effects of medications and interventions, and to allow the patient the greatest possible dignity and self-determination.

Table 1. **Common Regimens for Severe *Pneumocystis* Pneumonia†**

Anti-*Pneumocystis* Drug, Dose	Administration Route	Dosing Frequency
Trimethoprim/Sulfamethoxazole‡ 15 mg/kg/day TMP component	Oral or intravenous	Divided three times daily
Pentamidine‡ 4 mg/kg/day	Intravenous	Daily
Clindamycin/Primaquine 1800 mg/day Clindamycin 30 mg/day Primaquine	Clindamycin intravenous Primaquine oral	Clindamycin divided three times daily Primaquine daily
Trimetrexate/Leucovorin 30-45 mg/m^2/day Trimetrexate 20 mg/m^2/day Leucovorin	Intravenous	Trimetrexate daily Leucovorin divided four times daily
Atovaquone* 1500 mg/day	Oral	Divided twice daily with food

†Adjunctive corticosteroids are typically warranted in severe *Pneumocytis* pneumonia; common dosing: Prednisone (or equivalent) 40 mg twice daily days 1-5, 40 mg daily days 6-11, 20 mg daily days 12-21. ‡Standard initial therapies. *Atovaquone is generally used in mild to moderate *Pneumocystis* pneumonia.

Table 2. Select Investigations of Predictors of In-Hospital Mortality Following ICU Admission of HIV-Infected Patients

Authors Location	Patients Years	Predictors of Increased Mortality
Morris, et al[13] San Francisco, CA	58 HIV+ ICU admissions with PCP 1996-2001	Mechanical ventilation and/or pneumothorax No use of HAART ICU admission after 5th hospital day
Afessa and Green[7] Jacksonville, FL	169 HIV+ ICU admissions 1995-1999	Higher APACHE II score ICU transfers from another hospital ward (rather than emergency department) Lower $CD4^+$-cell count Lower serum albumin
Morris, et al.[24] San Francisco, CA	295 HIV+ ICU admissions 1996-1999	AIDS-related ICU admission diagnosis Lower serum albumin Higher APACHE II score PCP diagnosis Mechanical ventilation
Curtis, et al.[16] Multicenter – USA	237 HIV+ ICU admissions with PCP 1995-1997	Higher severity of illness Mechanical ventilation PCP diagnosis Prior use of PCP prophylaxis
Caslino, et al.[14] Paris, France	421 HIV+ ICU admissions 1990-1992	Higher WHO functional status Longer time since AIDS diagnosis Higher HIV disease stage Higher SAPS I score Mechanical ventilation Mechanical ventilation >10 days
Rosen, et al.[9] Multicenter – USA	63 HIV+ ICU admissions 1988-1990	Admission for pulmonary disorders Mechanical ventilation for pulmonary disorders

HIV+, human immunodeficiency virus infected. ICU, intensive care unit. PCP, *Pneumocystis carinii* pneumonia. APACHE, acute physiology and chronic health evaluation. SAPS, simplified acute physiology score.

REFERENCES

1.   Nakashima AK, Fleming PL. HIV/AIDS surveillance in the United States, 1981-2001. J Acquir Immune Defic Syndr 2003; 32 Suppl 1:S68-85.
2.   UNAIDS/WHO global AIDS statistics. AIDS Care 2003; 15:144.
3.   Torres RA, Barr M. Impact of combination therapy for HIV infection on inpatient census. N Engl J Med 1997; 336:1531-1532.
4.   Nuesch R, Geigy N, Schaedler E, et al. Effect of highly active antiretroviral therapy on hospitalization characteristics of HIV-infected patients. Eur J Clin Microbiol Infect Dis 2002; 21:684-687.
5.   Ghani AC, Donnelly CA, Anderson RM. Patterns of antiretroviral use in the United States of America: analysis of three observational databases. HIV Med 2003; 4:24-32.
6.   Palella FJ, Jr., Delaney KM, Moorman AC, et al. Declining morbidity and mortality among patients with advanced human immunodeficiency virus infection. HIV Outpatient Study Investigators. N Engl J Med 1998; 338:853-860.
7.   Afessa B, Green B. Clinical course, prognostic factors, and outcome prediction for HIV patients in the ICU. The PIP (Pulmonary complications, ICU support, and prognostic factors in hospitalized patients with HIV) study. Chest 2000; 118:138-145.
8.   De Palo VA, Millstein BH, Mayo PH, et al. Outcome of intensive care in patients with HIV infection. Chest 1995; 107:506-510.
9.   Rosen MJ, Clayton K, Schneider RF, et al. Intensive care of patients with HIV infection: utilization, critical illnesses, and outcomes. Pulmonary Complications of HIV Infection Study Group. Am J Respir Crit Care Med 1997; 155:67-71.
10.   Wachter RM, Luce JM, Turner J, et al. Intensive care of patients with the acquired immunodeficiency syndrome. Outcome and changing patterns of utilization. Am Rev Respir Dis 1986; 134:891-896.
11.   Wachter RM, Luce JM, Safrin S, et al. Cost and outcome of intensive care for patients with AIDS, Pneumocystis carinii pneumonia, and severe respiratory failure. Jama 1995; 273:230-235.
12.   Rosen MJ. Intensive care of patients with HIV infection. Semin Respir Infect 1999; 14:366-371.
13.   Morris A, Creasman J, Turner J, et al. Intensive care of human immunodeficiency virus-infected patients during the era of highly active antiretroviral therapy. Am J Respir Crit Care Med 2002; 166:262-267.
14.   Casalino E, Mendoza-Sassi G, Wolff M, et al. Predictors of short- and long-term survival in HIV-infected patients admitted to the ICU. Chest 1998; 113:421-429.
15.   Curtis JR, Bennett CL, Horner RD, et al. Variations in intensive care unit utilization for patients with human immunodeficiency virus-related Pneumocystis carinii pneumonia: importance of hospital characteristics and geographic location. Crit Care Med 1998; 26:668-675.
16.   Randall Curtis J, Yarnold PR, Schwartz DN, et al. Improvements in outcomes of acute respiratory failure for patients with human immunodeficiency virus-related Pneumocystis carinii pneumonia. Am J Respir Crit Care Med 2000; 162:393-398.

17.   Ledergerber B, Egger M, Opravil M, et al. Clinical progression and virological failure on highly active antiretroviral therapy in HIV-1 patients: a prospective cohort study. Swiss HIV Cohort Study. Lancet 1999; 353:863-868.
18.   Masur H. Critically Ill Immunosuppressed Host. In: Parrillo JE, Dellinger RP, eds. Critical Care Medicine: Principles of DIagnosis and Management in the Adult. St. Louis: Mosby, 2001; 1089-1108
19.   Dasgupta A, Okhuysen PC. Pharmacokinetic and other drug interactions in patients with AIDS. Ther Drug Monit 2001; 23:591-605.
20.   Soni N, Pozniak A. Continuing HIV therapy in the ICU. Crit Care 2001; 5:247-248.
21.   French M, Price P. Immune restoration disease in HIV patients: aberrant immune responses after antiretroviral therapy. J HIV Ther 2002; 7:46-51.
22.   Wendel KA, Alwood KS, Gachuhi R, et al. Paradoxical worsening of tuberculosis in HIV-infected persons. Chest 2001; 120:193-197.
23.   Wislez M, Bergot E, Antoine M, et al. Acute respiratory failure following HAART introduction in patients treated for Pneumocystis carinii pneumonia. Am J Respir Crit Care Med 2001; 164:847-851.
24.   Morris A, Wachter RM, Luce J, et al. Improved survival with highly active antiretroviral therapy in HIV- infected patients with severe Pneumocystis carinii pneumonia. Aids 2003; 17:73-80.
25.   Reilly JM, Cunnion RE, Anderson DW, et al. Frequency of myocarditis, left ventricular dysfunction and ventricular tachycardia in the acquired immune deficiency syndrome. Am J Cardiol 1988; 62:789-793.
26.   Berger JR, Sabet A. Infectious myelopathies. Semin Neurol 2002; 22:133-142.
27.   Ullrich R, Zeitz M, Heise W, et al. Small intestinal structure and function in patients infected with human immunodeficiency virus (HIV): evidence for HIV-induced enteropathy. Ann Intern Med 1989; 111:15-21.
28.   Call SA, Heudebert G, Saag M, et al. The changing etiology of chronic diarrhea in HIV-infected patients with CD4 cell counts less than 200 cells/mm3. Am J Gastroenterol 2000; 95:3142-3146.
29.   Bini EJ, Cohen J. Impact of protease inhibitors on the outcome of human immunodeficiency virus-infected patients with chronic diarrhea. Am J Gastroenterol 1999; 94:3553-3559.
30.   Nannini EC, Okhuysen PC. HIV1 and the gut in the era of highly active antiretroviral therapy. Curr Gastroenterol Rep 2002; 4:392-398.
31.   McArthur JC. Neurologic manifestations of AIDS. Medicine (Baltimore) 1987; 66:407-437.
32.   Suffredini AF, Ognibene FP, Lack EE, et al. Nonspecific interstitial pneumonitis: a common cause of pulmonary disease in the acquired immunodeficiency syndrome. Ann Intern Med 1987; 107:7-13.
33.   Beck JM, Rosen MJ, Peavy HH. Pulmonary complications of HIV infection. Report of the Fourth NHLBI Workshop. Am J Respir Crit Care Med 2001; 164:2120-2126.
34.   Singh F, Rudikoff D. HIV-Associated Pruritus: Etiolo and Management. Am J Clin Dermatol 2003; 4:177-188
35.   Koutsilieri E, Scheller C, Sopper S, et al. Psychiatric complications in human immunodeficiency virus infection. J Neurovirol 2002; 8 Suppl 2:129-133.
36.   Ratner L. Human immunodeficiency virus-associated autoimmune thrombocytopenic purpura: a review. Am J Med 1989; 86:194-198.

37.   Little RF, Wilson WH. Update on the Pathogenesis, Diagnosis, and Therapy of AIDS-related Lymphoma. Curr Infect Dis Rep 2003; 5:176-184.

38.   Scadden DT. AIDS-related malignancies. Annu Rev Med 2003; 54:285-303

39.   Masur H, Ognibene FP, Yarchoan R, et al. CD4 counts as predictors of opportunistic pneumonias in human immunodeficiency virus (HIV) infection. Ann Intern Med 1989; 111:223-231.

40.   Jones JL, Hanson DL, Dworkin MS, et al. Surveillance for AIDS-defining opportunistic illnesses, 1992-1997. MMWR CDC Surveill Summ 1999; 48:1-22.

41.   Wolff AJ, O'Donnell AE. Pulmonary manifestations of HIV infection in the era of highly active antiretroviral therapy. Chest 2001; 120:1888-1893.

42.   Montaner JS, Lawson LM, Levitt N, et al. Corticosteroids prevent early deterioration in patients with moderately severe Pneumocystis carinii pneumonia and the acquired immunodeficiency syndrome (AIDS). Ann Intern Med 1990; 113:14-20.

43.   Consensus statement on the use of corticosteroids as adjunctive therapy for pneumocystis pneumonia in the acquired immunodeficiency syndrome. The National Institutes of Health-University of California Expert Panel for Corticosteroids as Adjunctive Therapy for Pneumocystis Pneumonia. N Engl J Med 1990; 323:1500-1504.

44.   Limper AH, Offord KP, Smith TF, et al. Pneumocystis carinii pneumonia. Differences in lung parasite number and inflammation in patients with and without AIDS. Am Rev Respir Dis 1989; 140:1204-1209.

45.   Wright TW, Gigliotti F, Finkelstein JN, et al. Immune-mediated inflammation directly impairs pulmonary function, contributing to the pathogenesis of Pneumocystis carinii pneumonia. J Clin Invest 1999; 104:1307-1317.

46.   Kumar SD, Krieger BP. CD4 lymphocyte counts and mortality in AIDS patients requiring mechanical ventilator support due to Pneumocystis carinii pneumonia. Chest 1998; 113:430-433.

47.   Schein RM, Fischl MA, Pitchenik AE, et al. ICU survival of patients with the acquired immunodeficiency syndrome. Crit Care Med 1986; 14:1026-1027.

48.   Confalonieri M, Calderini E, Terraciano S, et al. Noninvasive ventilation for treating acute respiratory failure in AIDS patients with *Pneumocystis carinii* pneumonia. Intensive Care Med 2002; 28:1233-1238.

49.   Barnes PF, Bloch AB, Davidson PT, et al. Tuberculosis in patients with human immunodeficiency virus infection. N Engl J Med 1991; 324:1644-1650.

50.   Daley CL, Small PM, Schecter GF, et al. An outbreak of tuberculosis with accelerated progression among persons infected with the human immunodeficiency virus. An analysis using restriction-fragment-length polymorphisms. N Engl J Med 1992; 326:231-235.

51.   Rosenberg AL, Seneff MG, Atiyeh L, et al. The importance of bacterial sepsis in intensive care unit patients with acquired immunodeficiency syndrome: implications for future care in the age of increasing antiretroviral resistance. Crit Care Med 2001; 29:548-556.

52.   Brown MC, Crede WB. Predictive ability of acute physiology and chronic health evaluation II scoring applied to human immunodeficiency virus-positive patients. Crit Care Med 1995; 23:848-853.

53.   Bonarek M, Morlat P, Chene G, et al. Prognostic score of short-term survival in HIV-infected patients admitted to medical intensive care units. Int J STD AIDS 2001; 12:239-244.

54.   Alves C, Nicolas JM, Miro JM, et al. Reappraisal of the aetiology and prognostic factors of severe acute respiratory failure in HIV patients. Eur Respir J 2001; 17:87-93.

9

African Trypanosomiasis

Hayden T. White

Department of Medicine, University of the Witwatersrand, Johannesburg, South Africa

INTRODUCTION

African trypanosomiasis (sleeping sickness) results from infection due to organisms of the genus *trypanosoma* (order kinetoplastidae and family trypanosomatidae)(1). These organisms are digenetic parasites whose life cycle involves 2 hosts: a definitive mammalian host and an intermediate arthropod host (which is responsible for dissemination of the organism). They are classified into 2 groups; stercoraria and salivaria. **Stercoraria** comprises species that develop in the arthropod's hind gut (reduviid bug) and are therefore found in the bug's feces. This includes the causative agent of Chaga's disease or American trypanosomiasis, *Trypanosoma cruzi*. **Salivaria** include species that are found in the salivary glands of the tsetse fly (*Glossina* species) and are transmitted via inoculation into mammalian skin. The species infecting humans is known as *Trypanosoma brucei*. The brucei complex consists of the subspecies *gambiense* (west African trypanosomiasis, WAT) and *rhodesiense* (east African trypanosomiasis, EAT). Another subspecies, *T. brucei brucei* infects cattle, goats and sheep but not humans. While it has become useful to classify African sleeping sickness in terms of the 2 subspecies, it should be recognized that many different strains exist which are capable of infecting man (2).

MORPHOLOGY

The trypanosomiasis parasite is characterized by the presence of a free flagellum arising from the kinetoplast (an organelle containing DNA) and an undulating membrane that gives the organism motility (3,4). The organism exists in different stages in the various hosts. The epimastigote stage occurs in the insect vector (5,6). This stage is not infective to humans but transforms into the infective metacyclic trypanosome, which is considered the young form of trypomastigote. The trypomastigote is the mature form and is found in the blood of mammalian hosts. Polymorphic forms may be present and represent the brucei complex. If the species is monomorphic, then *T. cruzi* is diagnosed.

EPIDEMIOLOGY

The distribution of human trypanosomiasis follows that of the vector, the tsetse fly. Thirty-six countries in sub-Saharan Africa are considered to be endemic with some 50 million people at risk. Approximately 10 million km^2 are infested with tsetse, severely limiting the agricultural potential of these areas. Sleeping sickness has reached epidemic proportions in Angola, Uganda and the Sudan. Although 25000 new cases are reported annually, it is estimated that the true figure is closer to 300000 (7). During the 1960's the prevalence of sleeping sickness in most African countries had been reduced to < 0.1%. However, civil unrest over the past few decades and the subsequent decline in control programs has resulted in a marked increase in the prevalence of disease. Some areas have reported figures as high as 18%. Nowhere is this more evident then in the Democratic Republic of the Congo. The recent war has led to estimates that over 100000 people will die annually from trypanosomiasis. Another example is Angola, where 60% of the population in the north of the country has evidence of past or present infection. Civil wars and the breakdown of health services are the primary reasons for the resurgence of sleeping sickness in much of Africa.

VECTOR

The trypanosomes are transmitted by blood sucking flies of the genus *Glossina*. The fly may live for several weeks and is capable of transmitting the disease with each bite. The reproductive cycle of the parasite takes place in the midgut of the fly (6). The resulting metacyclic epimastigote enters the bite wound via the hypopharynx. They enter the blood stream where they begin to multiply by asexual binary fission, once every 6-8 hours. Later, some trypanosomes will enter the cerebrospinal fluid (CSF).

Flies are infected 18 to 35 days after feeding from an infected host. The vector for **WAT** is the riverine tsetse fly of the *palpalis* group. The fly is endemic throughout West and Central Africa. Its habitat is dense vegetation along rivers and forests. Humans are the preferred host but several species of animal may also be infected. Humans however, are the only known reservoir. The flies feed during the daylight hours and are attracted to dark skin. The incubation period tends to be long and asymptomatic carriers of the disease are common, increasing the risk of spread. **EAT** is transmitted by tsetse of the *morsitans* group. It is distributed from Uganda and Kenya in the north to Botswana in the south. The flies live in woodlands and thickets of the savanna. They are more inclined to bite animals then humans, who are incidental hosts. Bushbuck, hartebeest and cattle are the main reservoir. Man is usually infected while venturing into areas that the animals inhabit.

PATHOGENESIS

The pathogenesis is ultimately linked to the inability of the immune system to rid itself of the parasite. The African trypanosome is found in the extracellular compartment and survives in the mammalian host by periodically altering its surface antigenic coat, thereby aborting the developing immune response. The surface of the parasite is dominated by 2 glyoproteins, namely the variant surface glycoprotein (Vsg) of the blood stream stage and procyclin of the procyclic stage. The main function of the Vsg is to provide a protective coat that covers the entire surface of the parasite. The intention however, is not to avoid the immune system completely but to exploit it for the parasites benefit. The benefit of this system is that it leads to persistent infection due to the presence of a relatively constant and tolerable number of parasites rather then rapid killing of the host as would occur in the case of uncontrolled growth. Each parasite has the ability to display literally hundreds of different Vsg proteins and can evade the immune system indefinitely (8).

IMMUNOLOGY

Acquired immunity is antigen specific. Therefore, sleeping sickness is characterized by recurring parasitemias, with each new wave of parasites representing the selection of an immunologically distinct antigenic variant. During an infection, trypanosome variant surface glycoprotein (VSG) (9), determinants stimulate B cells through T cell dependent and T cell independent mechanisms. Polyclonal hypergammaglobulinemia, particularly increased immunoglobulin M is striking and a constant feature. However, little of the IgM produced is specific antitrypanosome antibody.

CNS DISEASE

Complex neuropathological changes are commonly found. Neurological complications occur more frequently with *T.b.gambiense*. The levels of several neurotransmitters may become significantly altered, especially in areas involved in sleep control. There can be marked increase in the levels of prostaglandin D_2, one of the ultimate sleep regulating substances (10). This may be responsible for several of the neurological manifestations including headache and somnolence. Pathologically, CNS involvement results in meningoencephalitis. Morular or Mott cells may be seen. These cells are plasmacytes with vacuolated cytoplasm and pyknotic nuclei that are thought to play a role in the production of immunoglobulin M (IgM).

CLINICAL PRESENTATION

Clinical manifestations of sleeping sickness are not pathognomonic and may vary. Broadly speaking, the disease progresses from local inoculation of the skin, to the hemolymphatic system, organ infiltration (heart and CNS) and ultimately death. *T.b. rhodesiense* tends to cause a fulminate disease with early CNS involvement, cardiac involvement and death within a few weeks. In contrast, *T.b. gambiense* infiltrates the lymph glands and the CNS later, and leads a chronic progressive course that may last months to years before death ensues.

A chancre may develop within the first 2 weeks at the site of the tsetse bite. Classically, it is painful, indurated and appears as a red papule, 2-5cm in diameter (11). It is not always present and tends to be found more often in non-Africans. Within 1-3 weeks of the bite, the parasite enters the blood stream. The invasion is accompanied by high fever, malaise, headaches, joint pains, tachycardia and hypoglycemia (glycolysis being the sole source of energy for the parasite). The fever may last for 1-7 days and then follows an intermittent course during which the patient feels well. An irregular circinate rash may appear. Further dermatological features include pruritus, hyperaesthesia (Kerandel's Sign) and edema of the face, hands and feet. Involvement of the eyes leads to interstitial keratitis and conjunctivitis. General lymphadenopathy follows as the disease progresses. Involvement of the spleen and liver leads to hepatosplenomegaly.

Death in patients with *T.b. rhodesiense* may occur before CNS involvement. Cardiac disease presents early with arrhythmia or cardiac insufficiency as a result of a pancarditis. Not surprisingly, the ECG tracing shows marked abnormalities. Pericardial effusions have also been documented.

The progression of disease is different for WAT and follows a more indolent course. Months to years may pass before the onset of clinical symptoms. Lymphadenopathy is a prominent feature of the disease. Characteristically, the supraclavicular and posterior cervical lymph nodes are enlarged (Winterbottom's Sign). This sign has also been associated with CNS involvement. Patients may present with irritability, insomnia, personality changes and lack of concentration long before the parasite is detectable in the CSF. Daytime somnolence and nocturnal insomnia is a prominent feature (12). Psychiatric disorders such as mania and delirium may intervene and patients with sleeping sickness have been found in psychiatric institutions. Patients may present with Parkinsonian like features with rigidity, shuffling gait and slurred speech. Eventually epilepsy intervenes and patients become severely disabled. The final stages of the disease include progressive dysfunction with patients being unrousable and subsequently are unable to eat or drink. This leads to progressive malnutrition, intercurrent infections and death.

Laboratory abnormalities that may be found include hemolytic anemia (Coomb's positive), abnormal liver function tests, thrombocytopenia, hypocomplementemia, crypglobulinemia, and abnormal clotting profiles indicative of disseminated intravascular coagulation. None of these are pathognomonic. Spurious hypoglycemia is sometimes observed due to the metabolism of glucose by the parasite. Recent studies have shown abnormalities with the hypothalamic-pituitary axis with resultant adrenal insufficiency, hypothyroidism and hypogonadism. The IgM in the serum and CSF is usually raised. CSF shows a mononuclear cell infiltrate and morular/Mott cells may be present.

DIAGNOSIS

It is important to make a definitive diagnosis of sleeping sickness. This is because the treatment varies depending on the strain of the offending parasite and organ system involved (i.e. CSF). Furthermore, the drugs used to treat the infection are extremely toxic (with an estimated mortality of 5-10% on treatment). The parasite can be isolated from blood, CSF, lymphnode and chancre aspirates. Serology is helpful but not diagnostic. The disease progression is nonspecific and may be confused with many others including, malaria, TB, brucellosis, syphilis, and viral encephalitis.

The parasite may be detected in peripheral blood. Both thick and thin smears, stained with either Giemsa or Wright's stains are required to confirm the diagnosis. *T.b. rhodesiense* is more easily diagnosed then *T.b. gambiense* as the later is less likely to be seen in peripheral blood. Blood is more commonly positive in the early stages of disease and should be examined on several occasions, as the parasitemia tends to occur intermittently. The two species cannot be differentiated by microscopy alone. Numerous techniques have been developed to increase the yield of specimens e.g. the use of an anion-exchanger DEAE-52 cellulose membrane and buffy coat examination (13). Both the haematocrit centrifugation and the quantitative buffy coat techniques have been evaluated in Uganda with varying degrees of success. For *T.b. gambiense*, the most dependable sites for recovering trypanosomes are aspiration of the chancre or a lymph node. Multiple specimens are usually required. Other, more laborious techniques include animal inoculation, which is the most sensitive test for *T.b. rhodesiense* and culture of the organism.

As the involvement of the CSF alters management and worsens prognosis, every patient with sleeping sickness should have a lumbar puncture. Demonstration of an increased white cell count (WCC) or elevate protein suggests CNS invasion. IgM is raised and morular cells may be seen. Parasites are not always detected, but the presence of WCC and protein abnormalities in the face of peripheral invasion is indicative of CNS disease. CSF IgM levels may remain elevated for long periods after effective treatment.

Serological testing has been used extensively. Patients with sleeping sickness produce a range of antibodies directed against variant surface glycoproteins as well as other antigens. Serological tests have the drawback of being unable to differentiate infected from exposed individuals. They are therefore important for population studies but are of limited use in diagnosing infected individuals. Newer tests are under investigation that may improve the diagnostic capabilities. The card agglutination test for trypanosomiasis (CATT) detects variant antigen types (VAT) of *T.b. gambiense*. The presence of VAT in the CSF correlates directly with the trypanosome infection of CSF. Unfortunately for *T.b rhodesiense*, the serodiagnosis relies on the detection of relatively small amounts of antibody against the invariant surface antigens, for which tests lack both sensitivity and specificity. A newer antigen detection system for *T.b. rhodesiense* uses a monoclonal antibody against a procyclic invariant antigen to detect the parasite in the peripheral blood (14). Field tests are underway in order to establish whether these serological techniques would be sufficient for parasite detection (15).

TREATMENT (table 1)

It is imperative that the diagnosis of HAT is accurate and the treatment is monitored for side effects and efficacy. Currently no distinction is made between infection and asymptomatic carriers; therefore if the parasite is found, the patient should be treated. The use of serology is less clear as one can be serologically positive in the absence of infection (16). These "serological suspects" are usually followed with sequential parasitological assays. Distinguishing between the two organisms is similarly important as this may alter management. Although it is impossible to distinguish them morphologically, DNA probes are available at a few research labs (17). Alternatively, the recommendations are to treat the patient according to the clinical presentation.

Early

Suramin. Suramin is a sulfonated naphthylamine polyanionic molecule. It was originally introduced in 1922 and is still the drug of choice for both *T.b. gambiense* and *T.b. rhodesiense*. A full course will cure virtually 100% of cases (18). It does not however cross the blood brain barrier so is of little use for treating CSF invasion (19). The mechanism of action is not known but suramin has been shown to inhibit various dehydrogenases and kinases including RNA polymerase and L-α-glycerophosphate. This drug is relatively slowly cidal, trypanosomes only disappearing from blood 12-36 hrs after injection.

Suramin is administered by slow intravenous injection in a 10% aqueous solution. The drug must be used within 30 minutes of reconstitution as it deteriorates in air. Suramin is a potentially toxic drug. Approximately 1 in 20000 people will have an idiosyncratic reaction that may include nausea, vomiting, seizures and shock. Other less significant and transient side effects include joint pain, fever, pruritus, urticaria, photophobia, conjunctivitis and paresthesias. Lab abnormalities include a transaminitis, raised urea and creatinine, and thrombocytopenia. Suramin is deposited in the renal tubules and may lead to albuminuria and renal failure. Albuminuria usually clears within a few weeks and therefore treatment should not be stopped. If however, the proteinuria worsens or casts appear, then an alternate treatment should be sought. Severe reactions including agranulocytosis, adrenal insufficiency, hepatitis and death have been reported. Most advocate the use of an initial test dose of 200 mg. Some would even consider giving a dose of steroids pre-treatment.

Pentamidine. Pentamidine is an aromatic diamidine, which was originally introduced in 1937. It is available in two preparations; the isothionate (1.74mg salt equivalent to 1 mg base) and the dimethane sulfonate (1.56mg salt containing 1mg base). While it is a potent inhibitor of nucleic acid synthesis by inhibiting S-adenosyl-1-methionine decarboxylase, the mechanism of action remains unclear. (20). Trials suggest that it has a prolonged action and slow rate of excretion. Following a single dose, volunteers were protected for as long as 295 days. Although small amounts can be detected in the CSF, it is not recommended for the treatment of late stage disease.

Pentamidine is recommended for the treatment of both EAT and WAT. While it is very effective in treating WAT the rates of cure are less than suramin for EAT. At one stage pentamidine was advocated for prophylaxis of persons at high risk and was given as an intramuscular injection every 3-6 months. The high cost and toxicity has led to the abandoning of this practice.

Minor side effects including hypotension, nausea, hypoglycemia, vomiting and tachycardia are temporary and should not interfere with treatment. The hypotension occurs secondary to histamine release but anaphylaxis is rare. The route of administration is usually via intramuscular injection although intravenous preparations are available. Injection sites tend to become very tender and sterile abscesses are not uncommon.

Diminazene aceturate. Like pentamidine, diminazene is an aromatic diamidine. Originally a veterinary compound, it has shown activity against EAT. There are no pharmacokinetic studies in humans although the half-life in sheep is 11-14 hours. Although not registered for human use it has been used extensively in endemic areas, as it is cheap and effective. The relapse rate is reported to be around 2-3% (21).

Late: CNS Invasion

Melarsoprol. Melarsoprol or Mel B was synthesized over half a century ago by Dr Freidheim. It was first made available in Africa in 1949. This drug will cure all stages of sleeping sickness. However, because of its toxicity, it is reserved for late disease. Melarsoprol contains 18.8% arsenic and is supplied in a 3.6% propylene solution. The mechanism of action is complex. It has been shown to irreversibly bind to trypanothione resulting in a compound called Mel T. Trypanothione represents over 80% of glutathione in trypanosomes and is essential for maintaining cellular resistance to oxidant stress. It may also interrupt glycolysis through the inhibition of the parasitic pyruvate kinase.

Despite being the drug of choice for late stage disease, the CSF penetration is poor with some authors reporting levels of near zero. Nevertheless, trypanosomes in the CSF are slower moving and fewer in number only 5-8 hours after the first injection. The dosing regimen is controversial. Numerous studies have reported various strategies from single dose therapy for early stage WAT to multiple dosing for late stage EAT. Certain generalizations are possible. For early stage WAT, a single dose of Mel B may be sufficient. In late stage disease, a cumulative dose of 30 - 50 mg will cure most patients.

In an attempt to simplify treatment with melarsoprol, a team of researchers in Angola studied the effects of a modified treatment protocol on side effects and outcomes in 767 patients with late stage WAT (22). The treatment schedule comprised 10 daily injections of 2.2 mg/kg of melarsoprol. Parasitological cure 24 h after treatment was 100% in both groups; there were six deaths (all due to encephalopathy) 30 days after treatment in each group. The number of patients with encephalopathic syndromes was also the same in each group. Skin reactions were more common with the new treatment, but all could be resolved by additional medication or withdrawal of treatment.

Side effects are common. They include, abdominal pain, nausea, albuminuria, hepatic dysfunction and exfoliative dermatitis. A Jarisch-Herxheimer type reaction may also occur following the lyses of trypanosomes. This can be prevented by the administration of a single dose of either suramin or pentamidine before initiating melarsoprol therapy. Extravisation during intravenous injection may lead to an intense local reaction. Polyneuropathy occurs in 10% patients. This does not respond to steroids and may progress to severe weakness. Although no formal evidence is available, some authors treat the neuropathy with regular thiamine injections.

Like all arsenicals, melarsoprol may cause severe CNS side effects. Reactive encephalopathy occurs in up to 18% of patients and in 1% leads to death. 2 types have been described i.e. a hemorrhagic encephalitis which is almost invariably fatal, and a reactive encephalopathy from which the large majority eventually recover. The onset is unpredictable. It is characterized by headache, tremor, and difficulty in speech, convulsions and coma (23). Encephalopathy will usually occur at the end of the first series of injections, during the interval between the first and second series or during the second series.

Treatment is controversial. The drug should be stopped and reinstituted slowly a few days later. Dimercaprol (a heavy metal chelator) and corticosteroids have been used to treat the encephalopathy. The impression of most clinicians is that dimercaprol is ineffective, which is not surprising given the hypothesis that the encephalopathy is immune mediated.

The treatment of the encephalopathy depends on the presentation (23);

1. On the first sign of CNS symptoms, the patient should be dripped with a 5% glucose solution. Mannitol 250 g/l can be administered and repeated if symptoms persist. The patient should receive 25 IU of ACTH or 50 mg prednisone immediately.

2. Convulsions should be treated with diazepam and epanutin. If persistent, repeat doses of mannitol may be given.

3. Subcutaneous injections of adrenalin may be given, provided there are no contraindications

4. General patient management strategies including fluid administration, airway management, level of consciousness etc. should be continued until patient improves.

If the patient survives, no sequelae are seen. There is some evidence that the prior administration of prednisone (with or without azathioprine) may decrease the incidence of encephalopathy. A large randomised trial comparing the concomitant use of prednisolone in patients with late stage Gambian trypanosomiasis demonstrated an overall reduction in the frequency of encephalopathy related deaths from 6.2 - 2.8% (24).

A-Difluoromethylornithine (DFMO)(Eflornithine). For the 6% of Gambian trypanosomiasis patients who relapse following melarsoprol therapy, eflornithine is remarkably effective. It is an irreversible inhibitor of ornithine decarboxylase, which is the key enzyme in the pathway leading to biosynthesis of polyamines, essential for proliferation of prokaryotic and eukaryotic cells (25). It is a suicide inhibitor, being a substrate of its target enzyme. The pharmokinectics have been well studied in humans. CSF penetration is excellent and contributes to the efficacy of eflornothine. In a large series from West Africa, gambiense disease refractory to arsenicals was treated with DFMO. The relapse rate was only 9% and is more common in patients with high CSF WBC and those treated with oral medication.

Although more convenient to administer, oral therapy should only be used in patients with poor venous access. It is safe but has a low efficacy and its duration of action is so short that frequent dosing is required. Although some advocate adding oral eflornithine for 3 weeks after the initial intravenous phase, this is probably of no added benefit. The minimum duration of treatment necessary for cure is unknown. Currently a 14-day course of intravenous therapy is recommended but trials are underway to evaluate the efficacy of a 7-day course. Documented side effects include, diarrhoea and anemia and more rarely, rash, seizures and thrombocytopenia (26).

Nifurtimox. Nifurtimox is a 5-nitrofuran that has been used since the mid 1970's to treat Chaga's disease. It inhibits the production of trypanothione reductase, leading to the production of superoxide. It has a short half-life and requires multiple daily dosing to maintain adequate levels.

Trials in WAT have been limited. Several small trials in the Sudan and Zaire on patients with CNS involvement have yielded mixed results.

Although the CSF appears to become sterile fairly rapidly, biochemical abnormalities continue to exist. Cure rates are said to range from 36% to 80%. Given the drugs questionable efficacy and substantial toxicity, it is currently only used in the treatment of melarsoprol resistant *T.b.gambiense* (27).

Nifurtimox is toxic at therapeutic doses. Along with the potential for inducing hemolytic anemia in patients with G6PD deficiency, there are numerous reports of severe CNS abnormalities. These range from a mild polyneuropathy to marked CNS symptoms of cerebellar disease, movement disorders and seizures. Most will resolve upon cessation of treatment but the long-term side effects are not known.

Future Therapies

Part of the difficulty with developing new agents is to identify targets in the parasitic metabolism that are not evident in the host. Trypanothione, a polyamine-containing analogue of glutathione, is unique to trypanosomatids. Inhibitors kill the organism by subverting the normal antioxidant role of trypanothione. Current drugs that are known to interfere with the synthesis include DFMO, nitrofurazone and some trivalent arsenicals.

A drug similar to DFMO, MDL73811 that is an inhibitor of S-adenosylmethionine decarboxylase has undergone animal experiments and was shown to be effective in eradicating *T.b. rhodesiense* infection in mice. Anticancer agents such as alkylphophocholines have had limited success in animal models of sleeping sickness. A recently developed diaminotriazine derivative (SIPI 1029) was found to be effective against resistant strains of *T.b. rhodesiense* in mice (28).

Table 1. Dosing regimens of the drugs used to treat African trypanosomiasis

Drug	Dosage	Therapy
Suramin	- 20 mg/kg ivi to max. 1 g - 200 mg test dose initially - dose given on day 1,3,7,14 and 21 to total of 5 grams	- slow ivi injection in 10% aqueous solution. - must be used within 30 min. of reconstitution
Pentamidine	- 3-4 mg/kg imi daily for 10 doses	- prepare in 3ml distilled water
Melarsoprol *	- 3.6 mg/kg ivi - 3-4 series of 4 injections separated by 1 week	- preparation includes deworming, short course antimalarial and single dose suramin/pentamidine - thiamine supplement - premedication with oral steroids - prednisolone (1mg/kg/d max. 40mg/d)
Eflornithine	- 100 mg/kg ivi 6hly for 14 days	- optional oral therapy for a further 2 weeks - main drawback is cost
Nifurtimox	- 10-15 mg/kg po for 60-90 days	- used mainly for relapse
Diminazine	- 5 mg/kg po every 2 days x 3 doses	- also available in im formulation

* An alternative regimen involves a 10-day non-interrupted course using 2.2 mg/kg/d of melarsoprol.

Intravenous injections of melarsoprol often lead to a severe phlebitis. This and the fact that cross infections with needles are common have led to the development of alternate routes of drug administration. Experimental work in the mouse model demonstrated good results with the topical melarsoprol gel therapy, alone or in combination with the nitrofuranes and the nitroimidazoles (29). Further studies are required, including the development of a transdermal delivery system, and pharmacokinetic and pharmacodynamic investigations before any clinical trials.

Combination therapy is becoming popular. Suramin was combined with various 5-nitroimidazole compounds and found to cure 100% of early stage sleeping sickness in mice. DFMO has been combined with suramin, bleomycin and melarsoprol and been shown to have improved activity against both early and late stage disease. Another promising combination is that of nitroimidazole and melarsoprol (30). Not only were there substantial cure rates for late stage, but the treatment course was shortened and lower levels of arsenicals were required, potentially reducing the incidence of encephalopathy.

VECTOR CONTROL AND SOCIAL PROGRAMS

During the early and mid 20th century, programs to limit the spread of the disease through vector control and population screening were relatively successful. Subsequently, the social and political unrest that has ravaged much of sub-Saharan Africa, has led to a marked resurgence of sleeping sickness (31). Early identification of infected individuals leads to better cure rates and lower costs and decreases the chance of transmitting the disease to others. It is therefore important to screen at risk populations in order to diagnose these cases early. As the treatment and prevention of trypanosomiasis is complicated and costly, efforts to control spread of the disease have centered on controlling the vector. The object is to reduce man/fly contact rather then to completely eradicate the fly. Various techniques have been employed with varying degrees of success including;

- Aerial and ground spraying with insecticides including DDT and more recently pyrethrins (deltamethrin).

- Targets and impregnated traps have reduced the spread of disease in several epidemics (Uganda, Congo, Cote d'Ivoire). The traps are inexpensive and nonpolluting. In a recent outbreak in the Busoga region of Uganda, 8000 pyramidal traps were distributed over an area of 2850 km^2. Within 3 months the fly density was reduced by 98% and the incidence of sleeping sickness by 80%.

- Live bait using insecticide sprays and dips or pour-ons for cattle in tsetse fly infested areas increase the chances of contact between fly and insecticide.

CONCLUSION

As the interaction between the vector and human populations is complex, it is imperative that sufficient social planning occurs before a plan of action is put into place. Much of the control mechanisms in the past involved the relocation of millions of inhabitants in an attempt to decrease fly/man interaction. This led to major disruptions to local economies and much suffering for the indigenous populations. Future plans will require consultation with the local inhabitants to improve participation. Combinations of the various prevention techniques are most likely to have the greatest impact on the spread of disease. These will have to be tailored to the needs of the community and local conditions. The WHO is already heavily involved in prevention strategies throughout Africa. Major funding from western governments is necessary for these projects to advance.

REFERENCES

1.   Hoare CA. *The trypanosomes of mammals. A Zoological Monograph.* Oxford: Blackwell, 1972.
2.   Molyneux DH, Pentreath V, Doua F. African Trypanosomiasis in Man. In *Manson's Tropical Diseases.* (20th edition), Cook GC, ed. London: WB Saunders Co, 1996.
3.   Smith DH, Pepin J, Stich A. Human African trypanosomiasis, an emerging public health crisis. Brit Med Bull 1998; 54: 341-355.
4.   Richards FF. The surface of the African trypanosome. J Protozool 1984; 31: 60-64.
5.   Wery M, Mulumba PM, Lambert PH, Kazumba L. Hematologic manifestations, diagnosis and immunopathology of African trypanosomiasis. Semin Hematol 1982; 19: 83-92.
6.   Vickerman K, Tetley L, Hendry KA, Turner CM. Biology of African trypanosomiasis in the tsetse fly. Biol Cell 1988; 64: 109-119.
7.   World Health Organisation. Planning overview of tropical disease control. Division of Control of Tropical Disease. Geneva: WHO, 1995.
8.   Barry JD, Graham SV, Fotheringham M, et al. VSG gene control and infectivity strategy of metacyclic stage *Trypanosoma brucei.* Mol Biochem Parasitol 1998; 91: 93-105.
9.   Donelson JE, Hill KL, El-Sayed NM. Multiple mechanisms of immune evasion in African trypanosomiasis. Mol Biochem Parasitol 1998; 91: 51-66.
10.   Pentreath UW, Rees K, Owolabi OA, Philip KA, Doua F. The somnogenic T lymphocyte suppression prostaglandin D2 is selectively elevated in

cerebrospinal fluid of advanced sleeping sickness patients. Trans R Soc Trop Med Hyg 1990; 84: 795-799.

11. McGovern TW, Williams W, Fitzpatrick JE, Cetron MS, Hepburn BC, Gentry RH. Cutaneous manifestations of African trypanosomiasis. Arch Dermatol 1995; 131: 1178-1182.

12. Murray HW, Pepin J, Nutman TB, et al. Recent advances. Tropical medicine. BMJ. 2000; 320: 490-494.

13. Cattand P, de Raadt P Laboratory diagnosis of trypanosomiasis. Clin Lab Med 1991; 11: 899-908.

14. Rebeski DE, Winger EM, Rogovic B, et al. Improved methods for the diagnosis of African trypanosomiasis. Mem Inst Oswaldo Cruz 1999; 94: 249-253.

15. Asonganyi T, Doua F, Kibona SN, et al. A multi-centre evaluation of the card indirect agglutination test for trypanosomiasis (Tryp Tect CIATT). Ann Trop Med Parasitol 1998; 92: 837-844.

16. Atouguia J, Costa J. Therapy of human trypanosomiasis: Current situation. Mem Inst Oswaldo Cruz 1999; 94: 221-224.

17. Truk P, Jammoneau V, Cuny G, Frezil J. Use of polymerase chain reaction in human African trypanosomiasis stage determination and follow-up. Bull WHO 1999; 77: 745-748.

18. Wang CC. Molecular mechanisms and therapeutic approaches to the treatment of African trypanosomiasis. Annu Rev Pharmacol Toxicol 1995; 35: 93-127.

19. Van Voorhis WC. Therapy and prophylaxis of systemic protozoan infections. Drugs 1990; 40: 176-202.

20. Sands M, Kron MA, Brown RB. Pentamidine: a review. Rev Infect Dis 1985; 7: 625-634.

21. Peregrin A, Mammen M. Pharmacology of diminazene. Acta Trop 1993; 54: 301-308.

22. Burri C, Nkunku S, Merolle A, Smith T, Blum J, Brun R. Efficacy of new, concise schedule for melarsoprol in treatment of sleeping sickness caused by *Trypanosoma brucei gambiense*: a randomized trial. Lancet 2000; 355: 1419-1425.

23. Pepin J, Milford F. African trypanosomiasis and drug induced encephalopathy: risk factors and pathogenesis. Trans R Soc Trop Med Hyg 1991; 85: 222-224.

24. Pepin J, Milord F, Guern Z, Mpia B, Ethier L, Mansinsa D. Trial of prednisone for prevention of melarsoprol induced encephalopathy in gambiense sleeping sickness. Lancet 1989; i: 1246-1250.

25. Milord F, Pepin J, Loko L, Ethier L, Mpia B. Efficacy and toxicity of eflornithine for the treatment of *Trypanosoma brucei gambiense* sleeping sickness. Lancet 1992; 340: 652-655.

26. Denise H, Barrett MP. Uptake and mode of action of drugs used against sleeping sickness. Biochem Pharmacol 2001; 61: 1-5.

27. Pepin J, Milford F, Meurice F, Ethier L, Loko L, Mpia B. High –dose nifurtimox for arseno-resistant *Trypanosoma brucei gambiense* sleeping sickness: an open trial in central Zaire. Trans R Soc Trop Med Hyg 1992; 86: 254-256.

28. Bacchi CJ, Vargas M, Rattendi D, Goldberg B, Zhou W. Antitrypanosomal activity of a new triazine derivative, SIPI 1029, *in vitro* and in model infections. Antimicrob Agents Chemother 1998; 42: 2718-2721.

29. Atouguia J, Jennings F, Murray M. Successful treatment of experimental murine *Trypanosoma brucei* infection with topical melarsoprol gel. Trans R Soc Trop Med Hyg. 1995; 89: 531-533.

30.   Jennings FW. Combination chemotherapy of CNS trypanosomiasis. Acta Trop 1993; 54: 205-213.
31.   Molyneux DH. Patterns of change in vector-borne diseases. Am Trop Med Parasitol 1997; 91: 827-839.

10

Parasitic Infections in the ICU

John Frean

National Institute for Communicable Diseases, National Health Laboratory Service, Johannnesburg, South Africa.

STRONGYLOIDIASIS

The nematode *Strongyloides stercoralis* is a unique pathogen. Its lifecycle allows opportunity for free-living existence; it is also one of very few helminth parasites that can reproduce entirely within a single host. It is the latter property that, under certain circumstances, results in serious or life-threatening disease.

Geographic Distribution

S. stercoralis is widely distributed in both tropics and temperate areas but is more common in warm, moist climates. Overall, one percent of the world's population is estimated to be infected (1). In some African countries the prevalence can be up 48% in the general population (2). A related parasite, *Strongyloides fülleborni*, causes human infections in Africa and New Guinea.

Life Cycle

The host range includes dogs and non-human primates, and animal handlers are sometimes at risk for infection (3). The infective stage for the human host is the filariform (L3) larva, which is able to penetrate intact skin, and pass via the lymphatics and venous system into the lungs. The larvae migrate into the alveoli, up the bronchial tree, and are eventually swallowed. There is a further moult in the small intestine and the newly-emerged adult female worms burrow into the mucosa. Egg production, which is parthenogenic, is about 20/day. Eggs hatch to release rhabitiform (L1, moulting to L2) larvae, which are normally passed in the stool to either enter the free-living cycle if environmental conditions are favourable, or become infective filariform larvae. A proportion of the rhabitiform larvae are able to proceed to the filariform stage in the gut, penetrating the gut wall or perianal skin to continue the life cycle within the host (termed autoinfection).

Pathogenesis and Clinical Presentation

The autoinfection process is responsible for the maintenance of covert long-term strongyloidiasis, and is central to the parasite's ability to cause serious or life-threatening disease if host immune mechanisms become defective. Strongyloidiasis has been diagnosed more than 40 years after presumed infection in ex-World War 2 prisoners-of-war in South East Asia (4). Low-grade infection is entirely asymptomatic in about a third of patients (1). Epigastric pain and watery diarrhoea are typical symptoms; there may also be nausea and vomiting. Weight loss may result from malabsorption and protein-losing enteropathy (5); this is the prime feature of *S. fülleborni* infections in children. Passage through the lungs can cause an eosinophilic pneumonitis (Löffler's syndrome). Some patients manifest a serpiginous, migratory urticarial eruption ('larva currens') related to the intradermal passage of migrating larvae. This may recur over prolonged periods of time. More common are transient non-migratory urticarial wheals around the waist and buttocks (5, 6).

Severe, Complicated Strongyloidiasis

Accelerated autoinfection gives rise to the dangerous complications of strongyloidiasis, namely hyperinfection and disseminated infection. The term 'disseminated' is sometimes restricted to ectopic location of adult worms (7), but clinically, it is simpler to distinguish between uncomplicated, and severe and complicated, infection (6). Cellular immunity is important to contain the infection, and a large number of underlying diseases or therapies that impair cellular defense mechanisms predispose to hyperinfection (5, 6). The most consistent association with hyperinfection is administration of corticosteroids. This has led to the theory that corticosteroid metabolism produces chemical mimics of parasite ecdysteroids (moulting hormones), which accelerate the development of the larvae in the host (8). This is speculative, however, and in some severe, complicated cases, immune depression preceded immunosuppressive therapy. Some of the conditions associated with hyperinfection include malignancies (lymphocytic and myeloid leukaemia, lymphomas, lung and gastric carcinoma); chronic renal disease (allografts, nephrotic syndrome, chronic glomerulonephritis); chronic infections (leprosy, tuberculosis); and a miscellaneous group (autoimmune disease, protein-calorie malnutrition, hypogammaglobulinaemia, achlorhydria) (6, 9). Although HTLV1 infection is particularly associated with hyperinfection, the condition does not commonly complicate HIV infection, although such cases have been described (10). Fatal strongyloidiasis in late pregnancy in a woman with no typical predisposing factors in South Africa, appears to be unique (11).

Clinical Features of Severe, Complicated Infection

The main targets of infection are the bowel, lungs and central nervous system. There is a spectrum of severity of presentation, but when there is massive invasion, mortality is high. Many symptoms and signs are non-specific and the diagnosis is frequently missed until late in the disease. Apart from abdominal pain, nausea, vomiting and diarrhoea, gastrointestinal syndromes include a sprue-like malabsorptive condition with steatorrhoea and protein-losing enteropathy (5); a pseudo-obstructive syndrome with signs of paralytic ileus; necrotising jejunitis, arteriomesenteric occlusion, small bowel obstruction, biliary obstruction due to papillary stenosis, aphthoid ulceration of the colon, and massive intestinal haemorrhage (6). The major worm burden is in the proximal jejunum, and penetration by filariform larvae is most pronounced in the distal small bowel and proximal large bowel. The majority of larvae pass into the intestinal lymphatics and then, via the thoracic duct, are carried to the lungs (12). Lung involvement may present with cough and dyspnoea, often with wheezing, and may progress to acute respiratory distress syndrome and fatal respiratory failure. Numerous filariform larvae are found in the lungs and sputum. Radiographic findings include diffuse pulmonary infiltrates, lobar consolidation, and occasionally abscesses. Filariform larvae may invade the central nervous system, producing a parasitic meningitis or cerebral vasculitis (5). Secondary bacterial infection commonly accompanies disseminated strongyloidiasis, resulting in bacteraemia, septicaemic shock, bacterial pneumonia, meningitis or brain abscess, peritonitis and endocarditis (9, 13). Enteric, usually Gram-negative, organisms are most commonly isolated (14), suggesting that injury to the bowel mucosa allows bacteria access to the bloodstream, and/or there is physical carriage of bowel flora on the invading larvae.

Diagnosis of strongylodiasis

On radiographic investigation, patients may present with Löffler's syndrome, with patchy, transient lung infiltrates. Extensive radiographic changes may be evident in complicated strongyloidiasis involving the lungs; radiographic features of obstruction, with loops of dilated bowel, may accompany severe hyperinfection (6).

On laboratory investigation the presence of a raised eosinophil count accompanying symptoms of diarrhoea, abdominal pain, and urticarial rashes is typical of strongyloidiasis, but laboratory confirmation requires the identification of the larvae. Culture of stool on a nutrient agar plate is more sensitive than the older Baermann technique (3) or examination of stool concentrates. *S. stercoralis* larvae must be distinguished from those of hookworm species hatching in a stool specimen which is not fresh. Laboratory personnel are at high risk for *S. stercoralis* infection and the agar plates must be sealed, and gloves must be worn during handling. In the case of severe, disseminated infection, there are usually numerous larvae, and sometimes adults and eggs in stool; filariform larvae, and occasionally adults, may be detected in sputum, bronchial lavage specimens, cerebrospinal fluid, blood, and other fluids or tissues. Raised IgE and eosinophil counts are not always present, especially in patients with overwhelming infection (7). Eosinophil counts may be suppressed by concurrent corticosteroid administration. The main value of serological tests (enzyme-linked immunosorbent assays) is in screening for evidence of low-level infection to identify candidates for more vigorous investigation (6).

Treatment

The traditional treatment for strongyloidiasis is thiabendazole, but it has erratic efficacy and is toxic. Safer and more reliable short-course agents are albendazole and ivermectin. Mebendazole is only effective in a long course (Table 1).

Table 1. Therapeutic approach to strongyloidiasis (6)

Clinical Situation	Suggested Regimens
Uncomplicated strongyloidiasis, initial treatment	albendazole 400 mg twice daily x 3 days ivermectin 200 µg/kg, single dose thiabendazole 25 mg/kg twice daily x 3 days
Severe, complicated or recurrent infection	albendazole 400 mg daily x 3 weeks ivermectin 200 µg/kg weekly x 4 weeks mebendazole 100 mg daily x 3 weeks
Suppressive therapy following failed eradication in immunocompromised patients	albendazole 400 mg x 1 day per month ivermectin 200 µg/kg x1day per month thiabendazole 25 mg/kg twice daily x 1 day per month

AMOEBIASIS

Molecular techniques have elucidated a long-standing controversy regarding the pathogenicity of *Entamoeba histolytica*. The former 'non-pathogenic strains' are now accepted as a separate, but morphologically indistinguishable, species, *E. dispar*. This old concept (15) was only relatively recently proved, with a substantial research contribution from South Africa (16). *E. histolytica* normally behaves as a commensal in the large bowel of humans. Invasive behaviour leads to active intestinal amoebiasis, with the potential for blood-borne spread to other organs.

Geographic Distribution

Amoebiasis is more common in developing countries than in industrialised nations (16, 17). About 10% of the world's population is infected with *Entamoeba* species; there are an estimated 40-50 million cases of invasive amoebiasis due to *E. histolytica* annually, with 40 000 to 110 000 deaths (18). High prevalences occur in parts of Central and South America, western and southern Africa (including KwaZulu-Natal Province, South Africa), India and the Far East (16). At high risk are those living in conditions of overcrowding and inadequate sanitation, and institutionalised, particularly mentally retarded, people (19).

Transmission Cycle

Infection is acquired by ingestion of cysts contaminating food or water, and occasionally, by mechanical transfer, e.g. by unsterile enema equipment (20). After excystation the trophozoites occupy the large bowel, colonising the lumen and mucosal surface, or living in the edges of ulcers if they are invasive, from where they may metastasise. They divide by binary fission, and form, in the intestinal lumen when stool passage is slow enough, chitin-walled cysts that are resistant to adverse environmental conditions.

Pathogenesis and Clinical Presentation

Only about 10% of asymptomatic carriers ever develop invasive disease. Pathogenic *E. histolytica* possess surface adhesins mediating attachment to colonic epithelium; other virulence factors are proteolytic enzymes, cell-free cytotoxins, contact-dependent cytolysis, and phagocytic ability (16). Poorly defined host factors, including nutritional status and physiological stress (e.g. pregnancy) are also involved (19). There is a wide spectrum of clinical presentation.

Acute Amoebic Colitis

Dysentery (passage of loose, grossly blood- and mucus-containing stools) of gradual onset (over 1 to 3 weeks), accompanied by cramps and tenesmus, are typical. There is usually a peripheral polymorphonuclear cell leukocytosis. True non-dysenteric colitis is probably extremely rare and its symptoms overlap considerably with irritable bowel syndrome (21). The differential diagnosis of amoebic dysentery includes bacterial infections (*Shigella, Salmonella, Campylobacter* spp., entero-invasive and –haemorrhagic *E. coli*) and non-infective conditions (inflammatory bowel disease, diverticulitis, ischaemic colitis).

Complications of Amoebic Colitis

Intestinal perforation due to transmural necrosis has variably subacute to acute presentations. Conservative management is sometimes appropriate (20); indications for surgery include peritonitis, air under the diaphragm, intra-abdominal abscess, colonic ischaemia, and severe haemorrhage (16). Fulminant colitis (toxic megacolon) with colonic dilatation, perforation and bacterial peritonitis, presents with diffuse abdominal pain, fever, hypotension, profuse bloody mucoid diarrhoea, and has a high mortality. Surgical management is partial or total colectomy with exteriorisation of the ends (22). Amoebomas are localised chronic amoebic infections of the large bowel, most common in the caecum; they may be mistaken for colon carcinoma, intestinal lymphomas, or enteric tuberculosis, if not associated with overt amoebic colitis (16, 20).

Extraintestinal amoebiasis

Amoebic liver abscess. This is the most common manifestation of extraintestinal amoebiasis. Presentation is often acute but may be insidious. Intestinal symptoms may or may not be present.

The typical amoebic liver abscess is single, located in the right lobe of the liver; hence the high frequency of right-sided abdominal or pleuritic pain. Patients with multiple abscesses may present more acutely (19). The liver may be enlarged and painful on palpation, but these signs are not invariable. Upward extension of the abscess may cause elevation of the diaphragm and atelectasis of the right lower lobe of the lung, sometimes with accompanying reactive pleural effusion, as distinct from pleuropulmonary amoebiasis (Table 2). Left liver lobe involvement may present with epigastric pain.

Diagnosis

Diagnostic testing. Adequate stool examination should precede barium studies to reduce interference with identification of amoebae. Radiological signs of colonic infection are mucosal oedema, haustral blunting, and ulceration, usually localised (19). Barium studies in acute infection sometimes carry a risk of perforation.

Table 2. Complications of amoebic liver abscesses and other extraintestinal amoebic disease (16, 20)

Syndrome	Main clinical features
Pleuropulmonary amoebiasis due to direct extension of liver abscess, or (rarely) haematogenous spread	Cough, chest pain, respiratory distress; abscess contents may be coughed up if hepatobronchial fistula occurs
Intraperitoneal rupture of liver abscess	Signs of acute septic peritonitis
Pericardial amoebiasis	Fever, abdominal pain, severe left-sided or retrosternal pain, pericardial rub, signs of acute congestive cardiac failure
Cerebral amoebiasis	Sudden onset, rapid progression; often fulminant course
Genital amoebiasis, usually due to fistulas from hepatic or rectocolonic infection	Lesions are painful, punched-out ulcers with profuse discharge
Cutaneous amoebiasis, by extension of abscess or intestinal infection into skin; also secondary infection of surgical wounds e.g. around colostomy site	Deep ulceration with necrotic base, and hyperplastic margins; painful. Sometimes extensive hyperplasia and minimal ulceration resembles verrucous carcinoma

Amoebomas have variable radiological appearances and may be difficult to distinguish from carcinoma. The plain chest radiograph in a patient with liver abscess commonly shows a raised right hemidiaphragm, right basal atelectasis, and a reactive pleural effusion. Fluid levels are not a feature of unruptured liver abscesses. Radioisotope scan of the liver is often the earliest imaging technique to be positive. Ultrasound typically reveals a hypoechoic round or oval lesion without significant wall echoes. Computed tomography (CT) and magnetic resonance imaging (MRI) also show liver abscesses well, but none of these techniques is specific for amoebic infection. Differential diagnosis includes pyogenic liver abscess and hepatoma.

Non-Specific Laboratory Findings. In mild colitis laboratory test results are generally normal. Most cases of liver abscess have a white cell count of more than 10 000/mm^3, and normochromic, normocytic anaemia is common. Alkaline phosphatase is raised in 70-75% of cases, whereas transaminases and bilirubin levels are abnormal in a much smaller proportion (20-50%) (17).

Detection of Amoebae. The mainstay of laboratory diagnosis of invasive amoebic colitis is still microscopic detection of haematophagous amoebae in stool; examination must be within 30 minutes of stool collection, and several specimens may be required. Microscopy of trichrome stained stool smears is ideal but its availability is limited. Amoebic culture, although more sensitive than microscopy is not available outside some specialised laboratories. Immunological tests for detection of *E. histolytica*-specific antigens or antibodies in stool, serum or saliva (23, 24) will ultimately displace microscopy in the routine laboratory; some such antigen tests are already commercially available (25). Colonoscopy with ulcer scrapings or biopsies may be indicated if the stool antigen test is negative or unavailable. Aspiration, ideally under ultrasound guidance, and culture of liver abscess material will help to distinguish bacterial from amoebic aetiology. Amoebic trophozoites are located in the abscess wall, rather than its centre, and are often very difficult to visualise microscopically.

Serological Tests. Outside endemic areas, in which up to a third or more of residents have persistent antibodies, positive amoebic serological tests (at presentation, or seroconversion) are very useful corroborative evidence of invasive amoebiasis, especially extraintestinal infection. Sensitivities of 90%-96% and 70%-85% for amoebic liver abscess and colitis, respectively, are mentioned (16, 17).

Treatment
Luminal carriage and invasive amoebiasis are treated differently (Table 3). Eradication of luminal amoebae after treatment of invasive amoebiasis is recommended to prevent relapses, as tissue amoebicides are not efficient at eliminating carriage. Luminal agents are variably available (e.g. in South Africa, none are sold). In this case, metronidazole (or tinidazole) is the only option for elimination of intestinal carriage, ideally with confirmation of cyst (or antigen) clearance. Because of high risk of reinfection and the lack of facilities for distinguishing *E. histolytica* from *E. dispar*, treatment of asymptomatic intestinal carriage in endemic areas is of questionable value.

Agents regarded as second-line or obsolete for treatment of invasive amoebiasis are dehydroemetine and emetine, chloroquine, and tetracycline.

Indications for aspiration of an amoebic liver abscess are to rule out a pyogenic abscess; to augment medical treatment, if there is no response after 72 hours; and to forestall imminent rupture, especially of a left lobe abscess.

Table 3. Treatment of *Entamoeba histolytica* infections (16)

Treatment Indications & Agents	Adult Dosage	Paediatric Dosage (mg/kg/day)
Asymptomatic intestinal carriage: luminal agents		
Diloxanide furoate	500 mg t.i.d x 10 days	20 in 3 doses x 10 days
Paromomycin	25-30 mg/kg/day x 7-10 days	25-30 in 3 doses x 7-10 days
Iodoquinol	650 mg t.i.d. x 20 days	20-40 in 3 doses x 20 days
Intestinal or extra-intestinal infection: tissue amoebicides		
Metronidazole followed by luminal agent (above)	750-800 mg t.i.d. x 10 days (intravenous: 1g stat, then 500 mg 6 hrly until oral dosing possible)	20 in 3 doses x 10 days
Tinidazole followed by luminal agent	800 mg t.i.d x 5 days or 2 g/day x 2-5 days	50-60 x 3-5 days

SCHISTOSOMIASIS

Several species of diecious flukes belonging to the genus *Schistosoma* cause human schistosomiasis, most importantly *S. mansoni*, *S. haematobium*, and *S. japonicum*. There is a wide spectrum of clinical disease, which sometimes includes acute, life-threatening complications. Clinical features in South Africa seem to be less dramatic than in other endemic areas (26).

Geographic Distribution
Africa is the continent most widely affected by the disease, caused mainly by *S. mansoni* and *S. haematobium*. *S. japonicum* is transmitted in several Far Eastern countries (China, Philippines, Indonesia). *S. intercalatum* and *S. mekongi* are geographically highly restricted to small areas of Africa and Southeast Asia, respectively. At present around 200 million people are infected, 85% of them in Africa (27).

Life Cycle

Each schistosome species utilises a different intermediate aquatic snail host, which becomes infected by the aquatic larval stage (miracidium) hatching from eggs excreted in urine or faeces of infected humans. A period of development in the body of the snail produces the cercariae, the infective stage for the human host, that are released from the body of the snail into the water for several weeks. Skin contact with water containing the cercaria is required for transmission. Humans are definitive hosts (i.e. they harbour the adult stages) of schistosome parasites. The final sites of predilection of the flukes are the veins of the bladder (*S. haematobium*) and the intestine (*S. mansoni* and *S. japonicum*). Immature stages may also be also responsible for disease, especially in immunologically naïve hosts, before maturity and egg laying occur. The number of flukes, and therefore the burden of eggs in a patient, is the major determinant of disease severity and outcome, but even a single pair of schistosomes, or a few eggs, in an ectopic site like the spinal cord or brain, may have disastrous consequences.

Clinical Features

It is convenient to divide schistosomiasis into 4 clinical stages: invasion, maturation, established infection, and late and complicated disease. The focus of this article is acute disease, so only the first two stages will be discussed; further information is available in several reviews (28 – 30).

Stage of Invasion. Cercarial dermatitis or 'swimmers itch' is papular skin irritation produced by the penetrating cercariae, more noticeable after heavy, compared to light, exposure. Previous exposure to cercariae may be sensitising, resulting in an exaggerated subsequent response. Passage of schistosomules through the lungs may result in transient coughing, again most likely to follow heavy exposure.

Stage of Maturation. At a variable interval (2 to 10 weeks) after exposure, an acute immunological reaction to the migrating schistosomules, or the maturing or ovipositing flukes, may occur. The clinical picture is one of acute serum sickness due to circulating immune complexes and other immunological phenomena, and is most likely to occur with *S. mansoni* or *S. japonicum* infections (31). In its more severe forms it is called acute toxaemic schistosomiasis or Katayama syndrome or fever; however, there is a spectrum of presentation from subclinical or mild to life threatening or fatal, the latter mainly a risk of *S. japonicum* infections. Although severe illness can result from reinfection, it is more likely to follow first exposure (32). It usually presents as fever, rigors, sweating, malaise, headache, and general myalgia Anorexia, nausea, abdominal discomfort, loose stools or diarrhoea, which may be dysenteric, are frequent. An urticarial rash and hepatosplenomegaly may be a feature. A dry cough and wheezing is frequently present. There is invariably a high eosinophil count. The acute pulmonary syndrome with systemic symptoms and diffuse nodular pulmonary infiltrates, typically related to *S. mansoni* exposure in Africa, has also been described in *S. haematobium* infections (33, 34). The pattern of cytokine secretion in peripheral blood mononuclear cells suggests that proinflammatory cytokines and low Th2 responses, together with immune complexes, are responsible. In contrast, patients with chronic schistosomiasis mansoni show a predominant Th2-type immune response, which may represent anti-inflammatory immune modulation (36).

Central nervous system infection is a relatively rare, but potentially devastating, feature of schistosomiasis. Cerebral schistosomiasis is most common in *S. japonicum* infections (2-5% of patients) (37) but spinal cord involvement complicates *S. mansoni* infections most frequently (28, 38), (although the same is claimed for *S. haematobium* in South Africa (26)). Acute cerebral schistosomiasis presents as a meningoencephalitis; in established or chronic disease, seizures, usually jacksonian type, are typical (28, 38). Spinal cord involvement should be presumed when there is a neurological lesion in the lower thoracic or upper lumbar region, microscopic or serological evidence of exposure to schistosomes, and other causes of myelitis are excluded (39). Two pathological patterns of spinal cord involvement have been distinguished: granulomas, and necrotic myelitis, the latter carrying a worse prognosis (40).

Diagnosis of Schistosomiasis
Parasitological Diagnosis. The 'first prize' in the diagnosis of schistosomiasis is direct visualisation of schistosome eggs in urine, faeces, or tissue biopsies (28). Various methods, including concentration techniques, are used in the routine diagnostic laboratory to detect schistosome eggs. For faeces, formol-ethyl acetate concentration or Kato-Katz cellophane thick faecal smears are typically employed. Centrifugation or filtration is used for urine examination. Rectal biopsy is relatively simple to perform; endoscopy of colon or bladder also provides mucosal samples for direct examination. Histological examination of tissue from ectopic sites is likewise may reveal presence of eggs. Post-treatment viability of eggs is important to evaluate.

Serological Diagnosis. Serological tests for schistosomiasis generally become positive relatively late in infection, remain positive for a long time after infection (confounding distinction between current infection and previous exposure), and sometimes give false positive results because of cross reactions (29). Nevertheless, serological tests may be useful in a previously non-exposed individual in whom eggs have not been detected. A variety of serological techniques and schistosome antigens are used. Some methods are claimed to distinguish acute from chronic disease, e.g. keyhole limpet haemocyanin (35), antibody subclass avidity (40), but these are not generally available. In suspected schistosomal myelopathy, antibodies in the CSF should be actively sought (41). Detection in serum and urine by ELISA of circulating cathodic and anodic antigens of adult schistosomes, is emerging as a promising technique, especially for survey work (28, 29).

Radiological diagnosis. The typical radiological feature of pulmonary involvement in acute schistosomiasis is diffuse nodular infiltration. High resolution computed tomography shows a characteristic ground-glass 'halo' around the nodules (42). Myelography, CT and MRI may usefully contribute to the diagnosis of brain or spinal cord infection. Ultrasound detection of periportal and peripancreatic lymphadenopathy has been reported in acute schistosomiasis mansoni (35). The main utility of ultrasound is screening for evidence of hepatosplenic and renal invovement in residents of endemic areas (28).

Treatment of acute schistosomiasis

The standard antischistosomal drug is praziquantel; alternatives that may be available in some parts of the world are listed in Table 4. Praziquantel is 60 – 90% effective, so a proportion of cases will require retreatment (30), particularly in acute infections as the drug is less effective against immature flukes than the adults (32). Artemether is active against immature schistosomes and in trials prevented acute disease (30). The clinical condition of patients may deteriorate transiently after treatment for acute schistosomiasis, presumably on an immunological basis (29).

There is still uncertainty around the use of steroids in addition to specific antischistosome drug treatment (28, 32, 43) in acute toxaemic disease. Neuroschistosomiasis should be treated early if reasonably suspected, even if not proven, because of potentially serious sequelae (44, 45). Adjunctive steroid treatment is generally recommended, although its efficacy has been questioned (38). Laminectomy may be required in acute paraplegia with spinal compression.

Table 4. Drug treatment of schistosomiasis (28)

Drug	Target species	Dosing information	Adverse effects
Praziquantel	All species	40-60 mg/kg, in 2 doses 4-6 hours apart	Nausea, abdominal discomfort, loose stools, headache, fatigue, skin eruption
Metriphonate	*Schistosoma haematobium*	10 mg/kg x 3 doses at 14 day intervals	Mild cholinergic symptoms: tremor, sweating, nausea, etc
Oxamniquine	*Schistosoma mansoni*	S. America, W. Africa 15 mg/kg, single dose North, East and southern Africa 20 mg/kg daily for 3 days	Dizziness, headache, drowsiness; occasionally abdominal discomfort, nausea, diarrhoea

PULMONARY EOSINOPHILIC SYNDROMES ASSOCIATED WITH PARASITIC DISEASES

There are a substantial number of conditions to consider in a patient with the combination of eosinophilia and lung involvement. Parasitic diseases in this category include larval nematode pneumonitis in ascariasis, hookworm infection and strongyloidiasis (Löffler's syndrome); acute schistosomiasis; tropical pulmonary eosinophilia related to filariasis; paragonimiasis; and echinococcosis. Non-parasitic diseases include certain fungal infections (bronchopulmonary aspergillosis, coccidioidomycosis, histoplasmosis); aspirin-sensitive asthma; drug-induced pneumonitis; and Churg-Strauss syndrome (vasculitis) (46). Acute pulmonary involvement in strongyloidiasis and schistosomiasis has been discussed earlier. Distinguishing features of tropical pulmonary eosinophilia and transient nematode pneumonitis (Löffler's syndrome) are shown in Table 5.

Investigation of the patient with unexplained eosinophilia starts with a full history (not forgetting past travel) and physical examination. A logical approach to laboratory investigations is advisable, in view of the diversity of potentially associated conditions. With regard to suspected parasitic

Table 5. Comparison between tropical eosinophilia and Löffler's syndrome (47).

Feature	Tropical eosinophilia	Löffler's syndrome
Associated infections	Lymphatic filariasis, dirofilariasis	Ascariasis, hookworm disease, strongyloidiasis
Eosinophilia	High (>3000 x 10^6/L)	Moderate (1000 – 3000 x 10^6/L)
Leukocyte count	Often high	Normal
Wheezing	Common	Rare
Systemic involvement	Common	Rare
Filarial serology	High titres despite no microfilaraemia	Negative
IgE levels	High (>10 000 ng/L)	Moderate
Duration	Recurrent	Transient
Response to diethylcarbamazine	Positive	Negative

infections, guided by history, clinical condition, and/or radiography or other imaging, a reasonable sequence of laboratory investigations is as follows:

- stool and urine microscopy, minimum of 3 examinations;
- rectal biopsy for schistosome eggs;
- *Schistosoma* antibody (or antigen, if available) test;
- thick and thin blood smear examination (night-time sample, if indicated);
- filarial antibody test and/or *Wuchereria* antigen test;
- serology and/or duodenal aspirate or stool culture for *Strongyloides;*
- hydatid and *Toxocara* serology.

Clearly, investigation and management need to be tailored to specific patients. If the patient is not acutely ill and all initial investigations prove negative, it may be reasonable to consider an empirical short course of an anthelmintic (eg albendazole), and investigate further should eosinophilia persist. On the other hand, life-threatening parasitic conditions such as eosinophilic meningitis, transverse myelitis, acute schistosomiasis japonica, disseminated strongyloidiasis, or hypersensitivity to hydatid antigens, require more aggressive investigation and treatment.

REFERENCES

1.   Stratton CW. Strongyloidiasis. Antimicrobics and Infectious Diseases Newsletter 2000; 18: 20-24.
2.   Petithory JC, Derouin F. AIDS and strongyloidisis in Africa. Lancet 1987; i: 921.

3. Despommier D, Gwadz R, Hotez PJ. *Parasitic Diseases*, 3rd edition. New York: Springer-Verlag, 1994.
4. Kennedy S, Campbell RM, Lawrence JE, Nichol GM, Rao DM. A case of severe *Strongyloides stercoralis* infection with jejeunal perforation in an Australian ex-prisoner-of-war. Med J Aust 1989; 150: 92-93.
5. Cook GC. *Strongyloides stercoralis* hyperinfection syndrome: how often is it missed? Q J Med 1987; 64: 625-629.
6. Grove DI. 'Human Strongyloidiasis.' In *Advances in Parasitology*, Vol 38. JR Baker, R Muller, D Rollinson, eds. San Diego: Harcourt Brace, 1996.
7. Scowden EB, Schaffner W, Stone WJ. Overwhelming strongyloidiasis. Medicine 1978; 57: 527-544.
8. Genta RM. Dysregulation of strongylodiasis: a new hypothesis. Clin Microbiol Rev 1992; 5: 345-355.
9. Gilman RH. '*Strongyloides* Infections.' In *Hunter's Tropical Medicine and Emerging Infectious Diseases*, 8th edition. GT Strickland, ed. Philadelphia: WB Saunders, 2000.
10. Cahill KM, Shevchuk M. Fulminant, systemic strongyloidiasis in AIDS. Ann Trop Med Parasitol 1996; 90: 313-318.
11. Poswa X, Kaka S, Frean J. Disseminated *Strongyloides stercoralis* infection in a pregnant patient. Proceedings of the 40th Annual Congress of the Federation of South African Societies of Pathology, Warmbaths, July 2000.
12. Haque AK, Schnadig V, Rubin SA, Smith JH. Pathogenesis of human strongyloidiasis: autopsy and quantitative parasitological analysis. Modern Pathol 1994; 7: 276-288.
13. Smallman LA, Young JA, Shortland-Webb WR, Carey MP, Michael J. *Strongyloides stercoralis* hyperinfestation syndrome with *Escherichia coli* meningitis: report of two cases. J Clin Pathol 1986; 39: 366-370.
14. Jain AK, Agarwal S, El-Sadr W. *Streptococcus bovis* meningitis associated with *Strongyloides stercoralis* colitis in a patient infected with human immunodeficiency virus. Clin Infect Dis 1994; 18: 253-4.
15. Brumpt E. Étude sommaire de l'*'Entamoeba dispar'* n. sp. Amibe à kystes quadrinuclées, parasite de l'homme. Bull Acad Med (Paris) 1925; 94: 943-952.
16. Jackson TFHG, Gathiram V. 'Amebiasis.' In *Hunter's Tropical Medicine and Emerging Infectious Diseases*, 8th edition. GT Strickland, ed. Philadelphia: WB Saunders, 2000.
17. Petri WA, Singh U. Diagnosis and management of amebiasis. Clin Infect Dis 1999; 29: 1117-1125.
18. WHO/PAHO/UNESCO report of a consultation of experts on amoebiasis. WHO Wkly Epidemiol Rep 1997; 72: 97-99.
19. Farthing MJG, Cevallos A-M, Kelly P. 'Intestinal Protozoa. The Sarcodina (amoebae).' In *Manson's Tropical Diseases*, 20th edition. GC Cook, ed. London: WB Saunders, 1996.
20. Ravdin JI. Amebiasis. Clin Infect Dis 1995; 20: 1453-1466.
21. Anand AC, Reddy PS, Saiprasad GS, Kher SK. Does non-dysenteric intestinal amoebiasis exist? Lancet 1997; 349: 89-92.
22. Aristizabal H, Acevedo J, Botero M. Fulminant amebic colitis. World J Surg 1991; 15: 216-221.
23. Abd-Alla MD, Wahib AA, Ravdin JI. Comparison of antigen-capture ELISA to stool-culture methods for the detection of asymptomatic *Entamoeba* species infection in Kafer Daoud, Egypt. Am J Trop Med Hyg 2000; 62: 579-582.
24. Abd-Alla MD, Jackson TF, Reddy S, Ravdin JI. Diagnosis of invasive amebiasis by enzyme-linked immunosorbent assay of saliva to detect amebic

lectin antigen and anti-lectin immuno-globulin G antibodies. J Clin Microbiol 2000; 38: 2344-2347.

25.   Haque R, Mollah NU,Ali IK, Alam K, Eubanks A, Lyerly D, Petri WA. Diagnosis of amebic liver abscess and intestinal infection with the *TechLab Entamoeba histolytica* II antigen dection and antibody tests. J Clin Microbiol 2000; 38: 3235-3239.

26.   Cooppan RM. Clinical features of schistosomiasis in the RSA. Continuing Medical Education 1989; 7: 162-169.

27.   Engels D, Chitsulo L, Montresor A, Savioli L. The global epidemiological situation of schistosomiasis and new approaches to control and research. Acta Trop 2002; 82: 139-146.

28.   Davis A. 'Schistosomiasis.' In *Manson's Tropical Diseases*, 20th edition. GC Cook, ed. London: WB Saunders, 1996.

29.   Strickland GT, Ramirez BL. 'Schistosomiasis.' In *Hunter's Tropical Medicine and Emerging Infectious Diseases*, 8th edition. GT Strickland, ed. Philadelphia: WB Saunders, 2000.

30.   Ross AGP, Bartley PB, Sleigh AC, et al. Schistosomiasis. N Engl J Med 2002; 346: 1212-1220.

31.   Evans AC, Martin DJ, Ginsburg BD. Katayama fever in scuba divers. S Afr Med J 1991; 79: 271-274.

32.   Rabello A. Acute human schistosomiasis mansoni. Mem Inst Oswaldo Cruz 1995; 90: 277-280.

33.   Cooke GS, Lalvani A, Gleeson FV, Conlon CP. Acute pulmonary schistosomiasis in travelers returning from Lake Malawi, sub-Saharan Africa. Clin Infect Dis 1999; 29: 836-839.

34.   Schwartz E, Rozenman J, Perelman M. Pulmonary manifestations of early schistosome infection among nonimmune travelers. Am J Med 2000; 109: 718-722.

35.   Rabello ALT, Garcia MMA, Pinto da Silva RA, Rocha RS, Chaves A, Katz N. Humoral immune responses in acute schistosomiasis mansoni: relation to morbidity. Clin Infect Dis 1995; 21: 608-615.

36.   de Jesus AR, Silva A, Santana LB, et al. Clinical and immunologic evaluation of 31 patients with acute schistosomiasis mansoni. J Infect Dis 2002; 185: 98-105.

37.   Watt G, Long GW, Ranoa CP, Adapon B, Fernando M, Cross JH. Praziquantel in treatment of cerebral schistosomiasis. Lancet 1986; ii: 529-532.

38.   Cosnett JE, van Dellen JR. Schistosomiasis (bilharzia) of the spinal cord: case reports and clinical profile. Q J Med 1986; 61: 1131-1139.

39.   Houpis J, Oexmann J, Martin J, et al. Acute schistosomiasis with transverse myelitis in American students returning from Kenya. MMWR 1984; 33: 445-447.

40.   Viana L de G, Rabello A, Katz N. Antibody subclass profile and avidity during acute and chronic human *Schistosoma mansoni* infection. Trans R Soc Trop Med Hyg 2001; 95: 550-556.

41.   Pammenter MD, Haribhai HC, Epstein SR, Rossouw EJ, Bhigjee AI, Bill PLA. The value of immunological approaches to the diagnosis of schistosomal myelopathy. Am J Trop Med Hyg 1991, 44: 329-335.

42.   Waldman AD, Day JH, Shaw P, Bryceson AD. Subacute pulmonary granulomatous schistosomiasis: high resolution CT appearances – another cause of the halo sign. Br J Radiol 2001; 74: 1052-1055.

43.   Lambertucci JR. Treatment of schistosomiasis: gathering stones together. Mem Inst Oswaldo Cruz 1995; 90: 161-164.

44.   Blanchard TJ, Milne LM, Pollok R, Cook GC. Early chemotherapy of imported neuroschistosomiasis. Lancet 1993; 341: 959.

45.   Scrimgeour EM. Schistosomiasis of the spinal cord – underdiagnosed in South Africa? S Afr Med J 1991; 79: 680.

46.   Philpott J, Keystone JS. Eosinophilia: an approach to the problem in the returning traveller. Travel Med Int 1987; 5: 51-56.

47.   Ottesen EA. 'The Filariases and Tropical Eosinophilia.' In KS Warren and AAF Mahmoud, eds. *Tropical and Geographic Medicine.* New York: McGraw-Hill, 1990.

11

Fungal infections in the ICU

W.D. Francois Venter and Ian M. Sanne
Reproductive Health Research Unit and Clinical HIV Research Unit, University of the Witwatersrand, Johannesburg, South Africa

INTRODUCTION

About 150 species of fungi are recognised as causing disease in humans. Many of these are nuisance infections, affecting the skin, mucosa and subcutaneous tissue. Very few fungi cause serious infections in humans, and before the advent of chemotherapy, immunosuppression, and human immunodeficiency virus (HIV) infection, systemic fungal infections were notably rare. In Africa, systemic fungal diseases have dramatically escalated with the HIV epidemic. Diseases treated as a small numbers of case studies collected over a century, are now part of the daily experiences for patients and clinicians.

Pneumocystis carinii shares morphologic and structural features consistent with protozoa and fungi. For many years, it was regarded as a protozoan by taxonomists. Clinicians were comfortable with the classification, as the disease caused by the organism does not respond to conventional antifungals, principally because the cell membrane lacks ergosterol, on which many antifungals work. However, subsequent analysis of pneumocystis RNA suggests substantial homology with other fungi, and the organism is widely regarded as such (1). *Pneumocystis carinii* is an important and common cause of disease in Africa, presenting overwhelmingly in HIV-positive people (2, 3).

Human mycoses are generally acquired from the environment. Human-to-human transmission is very rare, although sexual transmission of histoplasmosis is reported. Laboratory acquired infection is documented, and routine infection control procedures should be observed, with particular care taken with histoplasma and coccidioides. Humans are not natural hosts to all but a few isolated species, and infection represents no selective benefit to the species. Two yeasts, candida and cryptococcus, represent the vast majority of disease in the tropics (4-6).

CRYPTOCOCCAL INFECTION

Despite cryptococcus being isolated in 1894, the disease only became a clinical problem in the 1970's with the rise of widespread use of immunosuppressive therapies for transplantation, auto-immune diseases and malignancies. The advent of HIV infection produced an even greater explosion of the disease in the 1980's, and it has become the commonest disseminated fungal infection in this group. HIV infection in Africa brought a wave of cryptococcal disease, specifically cryptococcal meningitis. Blacks seem to be particularly at risk, and the disease appears to be especially prevalent in Southern Africa. Amphotericin B only suppresses the disease, without curing it.

The organism is ubiquitous, able to survive in a variety of ecologies, and cases have been described around the world. The organism is commonly found in soil and bird guano, although whether humans acquire it from these sources is unknown. The organism is rarely found in humans as a commensal, where it usually causes a transient infection.

Two important subgroups of *Cryptococcus neoforms* have been recognised – *gatii*, associated with eucalyptus trees and found in the tropics, but very rarely the cause of disseminated disease in AIDS patients, and *neoformans*, which accounts for almost all clinical disease (7-12).

Clinical presentation
The organism is inhaled through the lung in aerosolised form, infects a primary lymph node complex, and then disseminates systemically. The brain has an unexplained susceptibility to infection during systemic spread, but almost all organs have been involved (13).

Central nervous system infection. Central nervous system (CNS) infection is the commonest clinical manifestation of cryptococcal infection, usually presenting as meningoencephalitis. The disease rarely mimics classic bacterial meningitis and usually presents with fever, headaches, malaise, nausea, altered mentation, seizures, cranial nerve palsies, and blindness. Symptoms commonly occur over weeks and even months. Cryptococcosis is generally associated with a CD4 cell count below 100 cells/ml, in patients with AIDS. In these cases there may be a more rapid progression of symptoms than in those with other forms of immunosuppression. Cryptococcal meningitis is uncommon in HIV-infected children.

Diagnosis is easily made by examination of infected tissue or culture. CSF examination may be remarkably acellular, with a raised protein and low glucose. India ink examination is positive in three quarters of patients, but cryptococcal antigen detection in the CSF is extremely specific and sensitive (14-16).

Pulmonary cryptococcosis. The lung is the second commonest organ affected clinically, although most infections are asymptomatic. Subacute symptoms of cough, fever, and dyspnoea are the usual presenting features. "Allergic" pneumonitis is very uncommon, as is mass-related phenomena. The chest radiograph may demonstrate infiltrates, cavities, nodules, adenopathy, effusions, and masses. Localised pulmonary cryptococcosis is usual in HIV-negative patients, but uncommon in HIV-positive patients. With AIDS patients, concomitant infection with TB, *Pneumocystis carinii* and other respiratory pathogens is well-described, making diagnosis and successful treatment difficult (17, 18).

Other organs. The organism has been described to infect almost all organ systems. In AIDS patients, the isolation of cryptococcus in a non-central nervous system location usually means there is subclinical infection of the brain.

Treatment

The disease is rarely curable in AIDS patients, much like other diseases such as histoplasmosis and toxoplasmosis, in the absence of antiretroviral therapy. The goal therefore is to aggressively suppress infection until clinical signs have settled, and then to continue suppression.

Amphotericin B is generally indicated for all severe infections as initial therapy, and is probably best therapy for the initial therapy of all AIDS patients. Amphotericin B is very effective, despite a significant side-effect profile. Despite limited penetration beyond the blood-brain barrier, levels are adequate to achieve control of the fungal load in the meningeal space.

Azole therapy with itraconazole or fluconazole is very useful for treatment and prophylaxis of cryptococcal CNS infection. Azole therapy may be used as first-line therapy in stable patients without CNS involvement. Ketoconazole penetrates the CNS very poorly, and is only effective for infection not involving the meninges in HIV-negative patients. However, fluconazole is generally regarded as the azole of choice, due to its excellent side effect profile, especially compared to ketoconazole.

It is not clear whether combining an azole with amphotericin B is useful in the initial therapy of cryptococcal disease. Complications, such as hydrocephalus, may require surgical intervention (19, 20).

ICU management of this condition should always be considered. Even with AIDS patients, access to antiretrovirals may lead to a dramatic clinical improvement and expanded life span, if the episode of cryptococcus can be managed successfully. CNS cryptococcal disease can completely resolve if treated aggressively. Patients with altered levels of consciousness may require ventilation, while waiting to see the response to amphotericin B. In these cases, repeated spinal taps to reduce intracranial pressure should be considered, especially if prior taps have demonstrated improvement.

Occasionally, a patient may experience immune reconstitution syndrome on starting antiretroviral therapy, with sudden onset of neurological symptoms. A similar clinical response may occasionally be seen with the initiation of fungal infection. It is unclear how best to manage the syndrome – whether to interrupt antiretrovirals and give a course of antifungals, or whether continue antiretrovirals with antifungals, with or without steroids (21-24).

CANDIDIASIS

Candida is another pathogenic organism that has arisen as a major clinical problem in the late 20th century, as a consequence of medical progress. Unlike the other fungi, though, the rise of invasive tissue candidiasis is not linked to the rise in HIV, where it is an uncommon complication (25-28).

Over 200 species of candida are known, although only a handful is pathogenic in humans. *C. albicans* is the most common isolate in most cases. Isolation of other organisms has implications for drug susceptibility. Candida is the commonest cause of systemic fungaemia in the ICU. In the ICU, medical staff, and especially doctors are an important reservoir of candida infection (29).

Disseminated candidiasis carries a poor prognosis, partly due to delayed diagnosis and the underlying immunosuppression. Diagnosis can be very difficult, as blood, tissue and urine cultures can be negative despite wide dissemination. However, recovery of the organism from blood cultures has significantly improved in the last decade, through a variety of new laboratory methods. Autopsy series suggest that the disseminated candidiasis is seen in neonates, but is otherwise rare in children (30-32).

Candida can thrive in several situations. Bacterial competition in the gastrointestinal tract is frequently decreased through the use of broad-spectrum antibiotics, which have consistently been shown to be a major risk factor for candida. In one analysis, antibiotic use was the single biggest risk factor for candida fungaemia. Interestingly, use of certain antifungal azoles can select for different strains of candida (33). Neutropenia is a major risk factor, with dramatic increases in the candida colony load documented in multiple sites in the body. Cytotoxic therapy, immunosuppressive agents, corticosteroids, open surgical wounds, major trauma, and other immunosuppressive conditions may predispose to invasion, and the widespread use of antibiotics in this situation may promote increases in colony access to the body (34).

Entry is assumed to be via the gut, damaged integument or via vascular access devices. Wounds and burns have been commonly assumed to be risk factors for candidaemia, as integument colonization is common. Fungal endocarditis in intravenous drug addicts is well described, although endophthalmitis, osteoarthritis, and skin lesions also occur. Intravenous drug abusers are thought to be at risk, especially those using contaminated lemon juice to dissolve heroin. Controversy exists as to whether the access device allows colonization of a vascular channel, or whether intravenous solutions or hyperalimentation are colonized prior and during infusion (35).

Clinical features
The setting is the most important factor to consider when assessing the possibility of candidaemia. Patients in ICU's are at much higher risk than other patients – they are by definition severely ill and therefore immunocompromised, broad spectrum antibiotic use is common, prolonged hospital stays, and intravascular and other catheters are commonly in place (36). Presentation is usually of a fever without resolution on antibiotics. However, even fever is not invariable, as

immunosuppression may preclude a normal response. Endophthalmitis is an important clinical clue, but is not invariable (37).

Chest radiographic features of candida are uncommon, and multiple non-specific manifestations, from lymphadenopathy to interstitial patterns (38).

Treatment of systemic candidiasis

Most candida isolates are sensitive to amphotericin B, which is favoured as empiric therapy in severely ill patients with suspected fungaemia. However, *C. lusitania* is inherently resistant, and other species can develop resistance. Flucytosine is only available orally, but achieves high distributions into tissues, as well as crossing the blood-brain barrier. It is generally used in conjunction with amphotericin, and its efficacy in systemic candidiasis is not clear, although small studies suggest efficacy. The azoles are generally effective against candida, but there are differences between the different formulations, and a number of small case reports of clinical and microbiological resistance have been reported (39-43).

ASPERGILLOSIS

The term "aspergillosis" is used to describe a bewildering array of clinical diseases associated with exposure to the fungus, some of which may cause a patient to present to the ICU.

Aspergillus is principally a problem in Africa as a result of tuberculosis. An aspergilloma in the lung is probably the commonest serious manifestation of the disease in our setting. Damaged lungs, especially when cavitatory disease is present, may be colonized, and a mass of fungus develops. In a minority of patients, this may lead to haemoptysis, occasionally severe, and mortality is related to the extent of the haemoptysis and degree of underlying lung disease. Fungaemia is very rare. Radiographic appearance is characteristic, and the 'ball' may move within the cavity with different positioning. Management is difficult, and antifungal therapy is often of limited, if any, benefit. Surgery is advocated for severe or recurrent haemoptysis. Occasionally, a similar pathological process may occur in the sinuses.

Invasive aspergillus is associated with immunocompromise, especially with exposure to corticosteroids and neutropenia. It is a very rare AIDS-defining disease in HIV-positive patients. Invasive disease usually involves the lungs, although the brain, skin and other tissue involvement may occur. The diagnosis can be very difficult to make, as blood cultures are only occasionally positive. Amphotericin B is effective in very ill patients and itraconazole in those less seriously ill (41-47).

HISTOPLASMOSIS

The USA and Latin America carries the bulk of the burden of histoplasmosis, but the disease is recognised throughout the world. A handful of cases have been seen in Asia and Europe, but cases have been widely reported from central and southern Africa. *H.capsulatum* is known to occur naturally in caves within South Africa, Zimbabwe, and Tanzania and outbreaks amongst spelunkers is well described.

"African histoplasmosis" refers to disease caused by *Histoplasma capsulatum* var. *duboisii*, although var *capsulatum* also causes disease on the continent. In fact, *H. duboisii* is uncommon, with 250 cases described, with most cases confined to cases in central Africa, between 20°N and 10°S. *H. duboisii* tends to affect the lungs less, but causes more bone and mucocutaneous disease. It has distinct pathological features. Treatment is similar to *H. capsulatum*. Despite the dramatic impact of HIV in terms of other fungal infections, descriptions of AIDS-related disseminated *H. duboisii* infection have been few.

Several cases of laboratory-acquired histoplasmosis have been described (48-52).

Acute Pneumonitis
Epidemics of acute pneumonia due to inhalation of large concentrations of histoplasmosis can occur, and can occur in epidemics. Acute pneumonia is very difficult to diagnose unless it occurs in an epidemic, as the disease is very difficult to confirm by histology or culture, with a very wide clinical differential. Histoplasmosis should be suspected if several cases present one to two weeks after a common exposure, often associated with activity that disturbs dust contaminated with bird or bat droppings.

Clinically, the disease causes a range of symptoms, including fever, malaise, chest pain, cough, myalgias, rigors, nausea, loss of appetite, weight loss, and headache – the classic 'flu-like illness". Objective clinical signs beyond the signs of hypoxaemia are unusual. Very occasionally, disease spread to the pericardium can cause pericardial effusions, pericardial tamponade, and constrictive pericarditis. Extension to the mediastinum can cause life-threatening complications, especially if complicated by fibrosing mediastinitis. The chest radiograph is usually abnormal in severe cases, with bilateral pulmonary infiltrates, hilar adenopathy, and pleural effusions. Laboratory workup rarely shows more than a raised white cell count, and mild transaminitis. Biopsy and culture of tissue is almost always negative. Serological tests lack the sensitivity and specificity to make confident diagnoses.

Treatment is supportive, and most cases resolve spontaneously. Occasionally, respiratory support may be necessary. Pericardial and mediastinal disease may require surgery (48-50, 53-55).

Chronic Pulmonary Histoplasmosis
This disease occurs predominantly in Caucasians with chronic obstructive pulmonary disease. It may cause or aggravate hypoxia and cor pulmonale, and cavitation may cause life-threatening haemoptysis. The diagnosis is ideally made by culture of the sputum, but this can be very challenging, often requiring bronchoalveolar lavage. Treatment is with itraconazole or ketoconazole, for up to a year. Surgical resection does not help, as recurrence is invariable (53-55).

Disseminated Histoplasmosis
Disseminated histoplasmosis is potentially lethal, and has diverse clinical presentations, from weight loss to hepatosplenomegally to Addisonian crisis. The disease spreads via the lymphatics to regional lymph nodes, and to the reticuloendothelial system. Patients with immunosuppressive conditions form the largest group with this condition, with the rest made up of those at the extremes of age. Disseminated disease commonly affects the reticuoendothelial system in immunocompromised patients, but can also cause pulmonary, mucosal, adrenal, central nervous system, endocarditis, gastrointestinal, and other lesions. Chest radiographs are abnormal in half of the patients, with lymphadenopathy and a military infiltrate, with nodules smaller than those seen in acute pulmonary histoplasmosis. Bone marrow biopsy and culture is very useful in the diagnosis in 75% of cases, but biopsy of other appropriate tissue should be done as necessary. Bronchoalveolar lavage is useful in AIDS patients. Diagnosis should not be delayed while waiting for culture.

Treatment for non-meningeal and mild disease in non-immunosuppressed patients is adequately treated with azole therapy. Amphotericin B is necessary for more severe disease, and for those who are immunosuppression. Secondary prophylaxis is necessary for AIDS patients, with itraconazole (48,49,53,56-59).

Histoplasmosis and the ICU
Several clinical conditions precipitated by histoplasma may cause disease severe enough to warrant admission to an intensive care unit. Hypoxaemia secondary to acute pneumonitis may occasionally require respiratory support. Chronic pulmonary infections may precipitate haemoptysis severe enough to require intervention, and disseminated histoplasmosis can present with specific or multi-organ failure, particularly in AIDS patients (48-50, 53-55).

Histoplasmosis and AIDS
Fever and weight loss is the commonest manifestation of disseminated AIDS-related histoplasmosis, and the CD4 cell count is usually less than 100 cells/ml. 10% of cases present in a way similar to septic shock. 70-80% of cases have positive complement fixation or immunodiffusion tests. Urine and serum antigen testing is very sensitive and specific, but not yet generally available. Amphotericin B is the treatment of choice for patients with severe disease or meningeal involvement, but itraconazole is very effective for patients with less severe disease. Itraconazole is very effective as suppressive therapy. Ketoconazole is ineffective, with fluconazole less effective than itraconazole (56-59).

OTHER FUNGAL INFECTIONS IN THE TROPICS

Coccidioidomycosis
Coccidioidomycosis is almost entirely confined to people from the Americas, and tends to present as a sub-acute pneumonia, which usually resolves spontaneously. The fungus can cause disseminated disease, especially in immunosuppressed patients, and may uncommonly lead to ICU admission. Isolation of the organism from sputum or other tissue specimen culture, or from transbronchial biopsy, is diagnostic. Serology occasionally gives false-negatives, especially in HIV patients. Azole therapy is generally used as first line therapy, unless patients are severely ill, where amphotericin B is preferred (53, 60).

Penicillium marneffei

Penicillium is a rare infection, although more cases are rapidly arising in endemic areas in Asia, where it is AIDS-associated mycosis. Cases associated with brief travel to the area have been described. It presents as a disseminated infection, with weight loss, fever, anaemia, respiratory and cutaneous involvement, and the diagnosis is confirmed by microscopy or culture. Treatment is usually amphotericin B, followed by suppressive therapy with itraconazole (61, 62).

REFERENCES

1.   Edman JC, Kovacs JA, Masur H, Santi DV, Elwood HJ, Sogin ML. Ribosomal RNA sequence shows *Pneumocystis carinii* to be a member of the fungi. Nature 1988; 334: 519-522.
2.   Mahomed AG, Murray J, Klempman S, et al. *Pneumocystis carinii* pneumonia in HIV infected patients from South Africa. East Afr Med J 1999; 76: 80-84.
3.   McLeod DT, Neill P, Robertson VJ, et al, Pulmonary diseases in patients infected with the human immunodeficiency virus in Zimbabwe, Central Africa. Trans R Soc Trop Med Hyg 1989; 83: 694-697.
4.   Singh VR, Smith DK, Lawerence J, Kelly PC, Thomas AR, Spitz B, Sarosi GA., Coccidioidomycosis in patients infected with human immunodeficiency virus: review of 91 cases at a single institution. Clin Infect Dis 1996; 23: 563-568.
5.   Tesh RB, Schneidau JD. Primary cutaneous histoplasmosis. N Engl J Med 1966; 275:597-599
6.   Marques SA, Robles AM, Tortorano AM, Tuculet MA, Negroni R, Mendes RP. Mycoses associated with AIDS in the Third World. Med Mycol 2000; 38 (Suppl 1): 269-279.
7.   Matee MI, Matre R. Pathogenic isolates in meningitis patients in Dar Es Salaam, Tanzania. East Afr Med J 2001; 78: 458-460.
8.   Bergemann A, Karstaedt AS. The spectrum of meningitis in a population with high prevalence of HIV disease. Q J Med 1996; 89: 499-504.
9.   Hakim JG, Gangaidzo IT, Heyderman RS, et al. Impact of HIV infection on meningitis in Harare, Zimbabwe: a prospective study of 406 predominantly adult patients. AIDS 2000; 14: 1401-1407.
10.   Bogaerts J, Rouvroy D, Taelman H, Kagame A, Aziz MA, Swinne D, Verhaegen J. AIDS-associated cryptococcal meningitis in Rwanda (1983-1992): epidemiologic and diagnostic features. J Infect 1999; 39: 32-37.
11.   French N, Gray K, Watera C, et al. Cryptococcal infection in a cohort of HIV-1-infected Ugandan adults. AIDS 2002; 16: 1031-1038.
12.   Karstaedt AS, Crewe-Brown HH, Dromer F. Cryptococcal meningitis caused by *Cryptococcus neoformans* var *gatii*, serotype C, in AIDS patients in Soweto, South Africa. Med Mycol 2002; 40: 7-11.
13.   Aberg JA, Powderly WG. Cryptococcosis. Adv Pharmacol 1997; 37: 215-251.
14.   Moosa MY, Coovadia YM.; Cryptococcal meningitis in Durban, South Africa: a comparison of clinical features, laboratory findings, and outcome for human immunodeficiency virus (HIV)-positive and HIV-negative patients. Clin Infect Dis 1997; 24: 131-134.

15. Gumbo T, Kadzirange G, Mielke J, Gangaidzo IT, Hakim JG. *Cryptococcus neoformans* meningoencephalitis in African children with acquired immunodeficiency syndrome. Pediatr Infect Dis J 2002; 21: 54-56.

16. Mundy LM, Powderly WG. Invasive fungal infections: Cryptococcosis. Seminars in Respiratory and Critical Care Medicine 1997; 18: 249-257.

17. Batungwanayo J, Taelman H, Lucas S, et al. Pulmonary disease associated with the human immunodeficiency virus in Kigali, Rwanda. A fiberoptic bronchoscopic study of 111 cases of undetermined etiology. Am J Respir Crit Care Med 1994; 149: 1591-1596.

18. Aberg JA, Mundy LM, Powderly WG. Pulmonary cryptococcosis in patients without HIV infection. Chest 1999; 115: 734-740.

19. van der Horst CM, Saag MS, Cloud GA, et al. Treatment of cryptococcal meningitis associated with the acquired immunodeficiency syndrome. National Institute of Allergy and Infectious Diseases Mycoses Study Group and AIDS Clinical Trials Group. N Engl J Med 1997; 337: 15-21.

20. Ennis DM, Saag MS. Cryptococcal meningitis in AIDS. Hosp Pract (Off Ed) 1993; 28: 99-102, 105-107, 111-112.

21. Denning DW, Armstrong RW, Lewis BH, Stevens DA. Elevated cerebrospinal fluid pressures in patients with cryptococcal meningitis and acquired immunodeficiency syndrome. Am J Med 1991; 91: 267-272.

22. White M, Cirrincione C, Blevins A, Armstrong D. Cryptococcal meningitis: outcome in patients with AIDS and patients with neoplastic disease. J Infect Dis 1992; 165: 960-963.

23. Aberg JA. Reconstitution of immunity against opportunistic infections in the era of potent antiretroviral therapy. AIDS Clin Rev, 2000-01: 115-138.

24. Saag MS, Graybill RJ, Larsen RA, et al. Practice guidelines for the management of cryptococcal disease. Infectious Diseases Society of America. Clin Infect Dis 2000; 30: 710-718.

25. Fagon JY, Novara A, Stephan F, Girou E, Safar M. Mortality attributable to nosocomial infections in the ICU. Infect Control Hosp Epidemiol 1994; 15: 428-434.

26. Pittet D, Tarara D, Wenzel RP. Nosocomial bloodstream infection in critically ill patients. Excess length of stay, extra costs, and attributable mortality. JAMA 1994; 271: 1598-1601.

27. Jarvis WR. Epidemiology of nosocomial fungal infections, with emphasis on Candida species. Clin Infect Dis 1995; 20: 1526-1530.

28. Meunier F. Systemic candidal infections in neonates. Clin Infect Dis 1992; 15: 554.

29. Wingard JR. Importance of Candida species other than *C. albicans* as pathogens in oncology patients. Clin Infect Dis 1995; 20: 115-125.

30. Fraser VJ, Jones M, Dunkel J, et al. Candidemia in a tertiary care hospital: epidemiology, risk factors, and predictors of mortality. Clin Infect Dis 1992; 15: 414-421.

31. Martino P, Girmenia C, Venditti M, et al. Candida colonization and systemic infection in neutropenic patients. A retrospective study. Cancer 1989; 64: 2030-2034.

32. van de Wetering MD, Poole J, Friedland I, Caron HN. Bacteraemia in a paediatric oncology unit in South Africa. Med Pediatr Oncol 2001 37: 525-531.

33. Schwartz RS, Mackintosh FR, Halpern J, Schrier SL, Greenberg PL. Multivariate analysis of factors associated with outcome of treatment for adults with acute myelogenous leukemia. Cancer 1984; 54: 1672-1681.

34. Bodey GP. *Candidiasis: Pathogenesis, diagnosis and treatment.* New York: Raven Press, 1992.

35. Lecciones JA, Lee JW, Navarro EE, et al. Vascular catheter-associated fungemia in patients with cancer: analysis of 155 episodes. Clin Infect Dis 1992; 14: 875-883.

36. Smith C, Arregui LM, Promnitz DA, Feldman C. Septic shock in the Intensive Care Unit, Hillbrow Hospital, Johannesburg. S Afr Med J 1991; 80: 181-184.

37. Bodey GP. The emergence of fungi as major hospital pathogens. J Hosp Infect 1988; 11 (Suppl A): 411-426.

38. Buff SJ, McLelland R, Gallis HA, Matthay R, Putman CE. *Candida albicans* pneumonia: radiographic appearance. AJR 1982; 138: 645-648.

39. Hadfield TL, Smith MB, Winn RE, Rinaldi MG, Guerra C. Mycoses caused by *Candida lusitaniae*. Rev Infect Dis 1987; 9: 1006-1012.

40. Saiman L, Ludington E, Pfaller M, et al. Risk factors for candidemia in Neonatal Intensive Care Unit patients. The National Epidemiology of Mycosis Survey study group. Pediatr Infect Dis J 2000; 19: 319-324.

41. Soubani AO, Chandrasekar PH. The clinical spectrum of pulmonary aspergillosis. Chest 2002; 121: 1988-1999.

42. Jewkes J, Kay PH, Paneth M, Citron KM. Pulmonary aspergilloma: analysis of prognosis in relation to haemoptysis and survey of treatment. Thorax 1983; 38: 572-578.

43. Oren I, Goldstein N. Invasive pulmonary aspergillosis. Curr Opin Pulm Med 2002; 8: 195-200.

44. Denning DW, Stevens DA. Antifungal and surgical treatment of invasive aspergillosis: review of 2,121 published cases. Rev Infect Dis 1990; 12: 1147-1201.

45. Kallenbach J, Dusheiko J, Block CS, Bethlehem B, Koornhof HJ, Zwi S. Aspergillus pneumonia--a cluster of four cases in an intensive care unit. S Afr Med J 1977; 52: 919-923.

46. Benatar SR. Aspergillus infection in the Western Cape. S Afr Med J 1977; 51: 297-305.

47. Shirakusa T, Ueda H, Saito T, Matsuba K, Kouno J, Hirota N. Surgical treatment of pulmonary aspergilloma and Aspergillus empyema, Ann Thorac Surg 1989; 48: 779-782.

48. Houston S. Histoplasmosis and pulmonary involvement in the tropics. Thorax 1994; 49: 598-601.

49. Bradsher RW. Histoplamosis. Seminars in Respiratory and Critical Care Medicine 1997; 18:3: 259-264,

50. Craven SA, Benatar SR. Histoplasmosis in the Cape Province. A report of the second known outbreak. S Afr Med J 1979; 55: 89-92.

51. Valdez H, Salata RA. Bat-associated histoplasmosis in returning travelers: case presentation and description of a cluster. J Travel Med 1999; 6: 258-260.

52. Khalil MA, Hassan AW, Gugnani HC. African histoplasmosis: report of four cases from north-eastern Nigeria. Mycoses 1998; 41: 293-295.

53. Wheat LJ, Goldman M, Sarosi G. State-of-the-art review of pulmonary fungal infections. Semin Respir Infect 2002; 17: 158-181.

54. Goldman M, Johnson PC, Sarosi GA. Fungal pneumonias. The endemic mycoses. Clin Chest Med 1999; 20: 507-519.

55. Rippon, J.W. Medical Mycology, the pathogenic fungi and the pathogenic actinomycetes. 3rd ed. Philadelphia, PA: WB Saunders Company, 1988.

56. McKinsey DS. Histoplasmosis in AIDS: advances in management. AIDS Patient Care STDS 1998; 12: 775-781.

57.   Stevens DA. Diagnosis of fungal infections: current status. J Antimicrob Chemother 2002; 49 (Suppl 1): 11-19.
58.   Sarosi GA, Johnson PC. Disseminated histoplasmosis in patients infected with human immunodeficiency virus. Clin Infect Dis 1992; 14 (Suppl 1): S60-S67.
59.   Wheat J, Sarosi G, McKinsey D, et al. Practice guidelines for the management of patients with histoplasmosis. Infectious Diseases Society of America. Clin Infect Dis 2000; 30: 688-695.
60.   Feldman BS, Snyder LS. Primary pulmonary coccidioidomycosis. Semin Respir Infect 2001; 16: 231-237.
61.   Kaplan JE, Hu DJ, Holmes KK, Jaffe HW, Masur H, De Cock KM. Preventing opportunistic infections in human immunodeficiency virus-infected persons: implications for the developing world. Am J Trop Med Hyg 1996; 55: 1-11.
62.   Walsh TJ, Groll AH. Emerging fungal pathogens: evolving challenges to immunocompromised patients for the twenty-first century. Transpl Infect Dis 1999; 1: 247-261.

12

Endemic mycosis

Chadi A. Hage, Kenneth S. Knox, George A. Sarosi
Indiana University- School of Medicine, and Roudebush VA Medical Center, Indianapolis, IN 46202, USA.

INTRODUCTION

Histoplasmosis, blastomycosis, and coccidioidomycosis are the three major endemic fungi in North America. Although histoplasmosis is found in all continents except Antarctica, coccidioidomycosis in South America, and blastomycosis in Africa, only in North America are these illnesses common.

These three fungal diseases share many characteristics. The causative agents are mycelial soil organisms. Illness is acquired by inhaling aerosolized spores. In the infected host, the organisms change their form, a characteristic called dimorphism. *Histoplasma capsulatum* and *Blastomyces dermatitidis* convert to a yeast form at 37 degrees C (thermal dimorphism), whereas *Coccidioides immitis* converts in tissue to a spherule that replicates by forming endospores (tissue dimorphism).

The endemic areas are large. Most of the Midwest and South Central United States is endemic for both histoplasmosis [1] and blastomycosis [2], and a large area in the Southwest United States and an adjacent area of Mexico are endemic for coccidioidomycosis [3]. All three illnesses occur in normal hosts, although histoplasmosis and coccidioidomycosis are also major opportunistic mycoses in patients with depressed cell-mediated immunity, and especially in patients with acquired immunodeficiency syndrome (AIDS) [4],[5].

Histoplasmosis, blastomycosis and coccidioidomycosis are major T-cell opportunistic infections, as demonstrated by the very aggressive course seen in patients with AIDS, in whom T-cell deficiency is most severe.

HISTOPLASMOSIS

In the United States *H. capsulatum* var. *capsulatum* is responsible for the majority of the cases of histoplasmosis. *H. capsulatum* var. *duboisii* is predominantly found in south Africa and Europe (6). The spectrum of disease ranges from the asymptomatic acquisition of a positive histoplasmin skin test reaction to a rapidly fatal pulmonary or disseminated illness. It is the balance between the net immune status of the subject and the load of the infecting inoculum that determines the severity of the illness. During the past two decades, histoplasmosis has emerged as a common opportunistic infection in patients with AIDS especially those residing in endemic areas. Most of our current knowledge about this disease is derived from outbreak investigation in the Midwestern United States.

The Pathogen and Epidemiology

Histoplasma is a thermally dimorphic fungus. It has two forms; a mycelia phase and a yeast phase. Mycelia are the forms found in the environment, they are considered to be the infectious form. Mycelia display macro-and micro-conidia. Yeasts are what are found in the infected individuals. In the laboratory these two forms are inter-convertible by altering the temperatures and nutrients of the growth medium. The disease is highly endemic in the Midwestern and southeastern parts of the US. It is estimated that 50-80% of people living in the Ohio and Mississippi river valleys have evidence of remote infection with histoplasma [7]

Human and animal infections occurred after inhalation of aerosolized micro-conidia, the infecting particle, which can reach the alveolar space due to its very small size of 2 to 5 mm. Extensive skin test surveys suggest that as many as 50 million people in the United States have been infected by *H. capsulatum* and that there are up to 500,000 new infections yearly [8].

For decades most outbreaks have occurred in urban settings. Most are associated with construction projects that disturbed contaminated soil. The most recent (and largest ever) outbreak occurred in Indianapolis, Indiana [9] , associated with downtown construction of a swimming pool complex.

Mini-outbreaks also still occur. Activities such as cutting up fallen trees or cleaning large bushes have been linked to smaller outbreaks. Histoplasmosis occurs in 2 to 5% of patients with AIDS who reside in endemic areas and up to 25% in selected areas during periods of outbreaks [10].

Pathogenesis. The lung is the portal of entry in almost every case. Due to their small size, microconidia can reach the alveolar space where they convert to the yeast form, a key step in the pathogenesis. The initial tissue response to the organism is predominantly neutrophilic, followed by an increase in alveolar macrophages [11]. After being phagocytosed, yeast survive and actually proliferate inside the macrophages, which serve then as a carrier throughout the reticuloendothelial system. Shortly after infection the yeast forms can be identified in the mediastinal lymph nodes. The fungus then gains access to the circulation. It is likely that transient self-limited fungemia occurs in most, if not all, patients. Later the yeast disseminates throughout the body to establish foci of infection in many organs, such as the liver, spleen and the adrenal glands [5].

About two weeks after infection, specific cellular immunity begins to develop. Activated lymphocytes secrete cytokines that stimulate macrophages in an attempt to boost their fungicidal activity. In mice, interleukin-12 is an important signal, leading to increased interferon-gamma production that confers protection against primary infection [12], [13], [14]. Tumor necrosis factor-alpha seems to be an important element in this scheme. Inhibition of TNF-α has been shown to alter the adaptive immune response to histoplasma infection and may predispose patients disseminated infection [12], [15], [16].

With the advent of T-lymphocyte-mediated cellular immunity, fungal replication is checked and granuloma formation begins. Healing of these lesions is accompanied by peripheral fibrosis. Central areas of encapsulated, necrotic material frequently calcify. These calcified foci manifest on chest roentgenogram as single or multiple calcified nodules. Calcified lesions are often seen in the liver and the spleen [17].

If the cell-mediated immune response is poor, the yeast continues to multiply. More macrophages are recruited, which in turn become parasitized and eventually disrupted, perpetuating the cycle [5]. A severe systemic illness develops, which invariably leads to death unless treated promptly and aggressively.

Clinical Manifestations. Most normal persons who are infected by the fungus remain asymptomatic. When present, symptoms vary widely from brief periods of malaise to severe, life-threatening illness. It is classically a flu-like illness with abrupt onset of fever, chills, and substernal chest discomfort. A harsh, nonproductive cough develops along with headache, arthralgias, and myalgias.

In immunocompetent individuals severe pulmonary illness develops rarely, only when the infective dose is unusually high. It may progress rapidly to the acute respiratory distress syndrome, and may lead to death from respiratory insufficiency if not treated promptly [1].

Immunocompromised individuals are more likely to progress to disseminated disease even after an infection with a smaller inoculum. The vast majority of progressive disseminated histoplasmosis is seen in patients with advanced AIDS with CD4 counts below 200 cells/mL [10].

Progressive Disseminated Histoplasmosis

Most patients with progressive disseminated Histoplasmosis present with fever, chills, weight loss, cough and progressive dyspnea [18]. On physical examination, patients are febrile and acutely ill. Hepatosplenomegaly may be present. Laboratory evaluation shows anemia, leukopenia, and thrombocytopenia. In extremely ill patients the syndrome of disseminated intravascular coagulation may be seen. Up to 20% of patients present with severe septic shock, respiratory failure and progress to multi-organ failure. This syndrome represents an advanced stage of the illness and is usually seen when diagnosis and appropriate therapy were delayed. Mortality is very high with this scenario.

The chest roentgenographic findings are variable, ranging from normal to diffusely abnormal, with reticulonodular pattern being the most frequently reported finding. Pleural effusions are rarely seen [19]. The radiographic findings are very similar to those seen with *Pneumocystis carinii* pneumonia [4] (Figure 1).

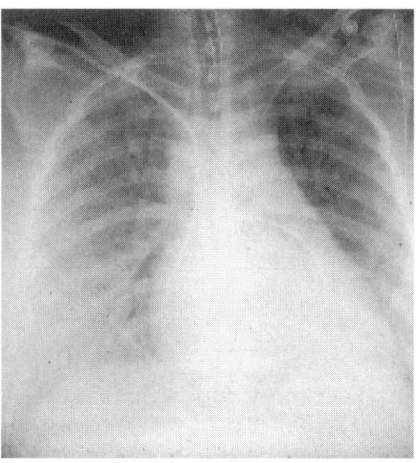

Figure 1
Chest X-ray of a HIV + patient with progress
disseminated Histoplasmosis.

Peripheral blood smear may show phagocytoses yeast in some of patients with severe disease. Biopsy material from the bone marrow and other involved tissue shows collections of macrophages full of intracellular yeast or, in the most severe instances, widespread necrosis with large numbers of organisms lying loose in the extracellular debris. There is little, if any, evidence of granuloma formation [20] . Virtually all patients with this form of the illness have some degree of T-cell defect. Before the modern era of widespread use of cytotoxic agents and glucocorticoids, many patients had underlying Hodgkin's disease, a well-known example of naturally occurring T-cell immune deficiency [5], [4].

The most severe form of progressive disseminated histoplasmosis (PDH) occurs in patients with AIDS with profound T-cell dysfunction [21]. In fact, most cases of PDH now occur in AIDS patients, and most occur in highly endemic areas [22], [23]. Newer biological therapies for Rheumatoid Arthritis and other auto-immune disorders have added a new pool of immunosuppressed patients at risk for tuberculosis and also histoplasmosis and other T-cell opportunistic fungal infections.

In most instances, exposure of an immunosuppressed person to an infected aerosol is the antecedent event preceding PDH. In the recent large outbreak in Indianapolis, most patients who were immunocompromised when they developed primary histoplasmosis progressed to PDH [23]. In particular, patients with AIDS nearly always progressed to PDH [10].

In other cases, the onset of PDH is temporally related to intense immunosuppression, most commonly progression of AIDS or therapy with high doses of glucocorticoids. In some of these cases, reactivation of dormant histoplasmosis may be the mechanism of infection [4]. Patients with AIDS who develop PDH after long residence in New York City or San Francisco are clear examples of such endogenous reactivation because primary infections are never seen in these cities.

Some patients with PDH present with a subacute to chronic illness. They may have a chronic wasting disease with anorexia, weight loss, and low-grade fever. Mucosal and mucocutaneous junction ulcers may occur in the mouth, oral pharynx, rectum, and glans penis. Adrenal involvement may cause Addison's disease [24]. Biopsy material from involved tissues shows well-formed epithelioid granulomas, and only a diligent search will reveal rare organisms. Demonstration of organisms almost always requires special stains [5], [20]. The disease may be systemic or involve only one organ. This more chronic form of PDH generally occurs in patients who are less immunosuppressed than the patients who develop more fulminant PDH.

Central nervous system histoplasmosis is rare and may present as chronic meningitis or intracranial histoplasmoma [25]. Endocarditis can also occur, involving either the aortic or mitral valves. Vegetations are usually large, and emboli are common. Endocarditis may occur on prosthetic or previously normal valves. Recently, histoplasma involvement of abdominal aortic aneurysms has been reported in a few patients with the chronic form of PDH [26].

Diagnosis

The gold standard of diagnosis is culture of the fungus from biologic material. Cultures are time consuming and one cannot wait for them in the management of cases with severe PDH. Delay in diagnosis while awaiting results of fungal cultures may lead to a fatal outcome in more severe cases Isolation of *H. capsulatum* may occur within 1 week in a minority of patients but usually takes several weeks.

Serologic studies of Histoplasmosis are seldom useful in the management of PDH. Positive serology does not predict development of PDH in patients undergoing bone marrow transplantation [27]. The complement fixation test is negative in up to 50% of patients with PDH. Immunodiffusion fails to identify up to 50% of patients with acute histoplasmosis and usually does not reach maximum positivity for 4 to 6 weeks after exposure [28]. Their main drawbacks are imperfect sensitivity and lack of timeliness. Several weeks must pass before they become positive. By that time, most patients have either recovered or have required other more invasive methods of diagnosis because of rapidly worsening disease

There are two ways to make a rapid diagnosis of PDH, sampling and examination of likely infected tissue with the use of special stains and the use of the ultrasensitive assay for fungal antigens.

Mycologic Studies. Bronchoscopy is an important diagnostic tool, especially for PDH. Specimens obtained from bronchoscopy have a high but poorly defined yield in severe primary histoplasmosis with progressive ARDS and especially in PDH in AIDS, when diffuse infiltrates are one of the clinical features. In highly selected series the diagnostic yield of bronchoscopy for diagnosis of histoplasmosis in an endemic area is about 60% in patients with infiltrates [29]. In a strictly AIDS population in Indianapolis, Indiana, fungal stains performed on bronchoalveolar lavage fluid provided a rapid diagnosis in 70% of patients; diagnostic yield increased to 89% when culture results were included. In that series, 22% of patients had co-infections or alternative diagnoses that were detected by BAL and would not have been detected if Histoplasma antigen testing had been the sole diagnostic test [30]. Use of cytologic examination without special fungal staining (silver, PAS) may explain the lower yield of BAL reported in series from non-endemic areas [31]. It is likely best to do a battery of stains including a silver stain. Although transbronchial biopsy is not mandatory at the time of bronchoscopy and BAL, histopathology does appear to enhance the diagnostic yield [31]. The fungus is difficult to see on standard hematoxylin and eosin stains; special stains (usually a silver stain) are needed. Special stains are particularly important when well-formed granulomas are present because of the paucity of organisms in such cases.

In patients with suspected PDH, sampling of the reticuloendothelial system is often effective for diagnosis. Bone marrow biopsy is likely the best and safest method [20] . In heavily parasitized samples, a direct smear of the bone marrow, stained with a supravital stain such as the Giemsa stain, usually gives a rapid diagnosis (Figure 2). On permanent histologic sections, the fungus is difficult to see on standard slides prepared with hematoxylin and eosin stain. It is best to go directly to special stains, usually one of several modifications of the silver stain or the PAS stain.

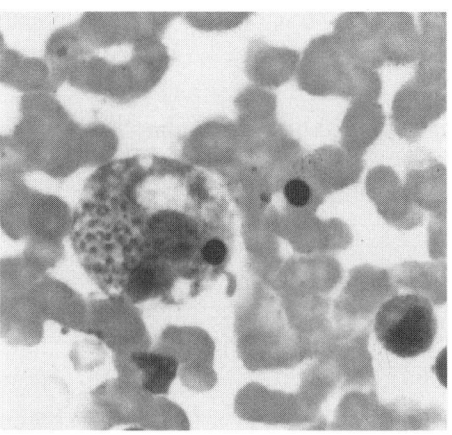

Figure 2
Intraphagocytic yeasts seen on a bone marrow aspirate.

Recently, the role of blood cultures in diagnosis of PDH has expanded. The lysis-centrifugation system increases sensitivity. In AIDS patients with PDH, the density of organisms is higher than in other immunosuppressed patients, and blood cultures are particularly useful, yielding a diagnosis in up to 90% of cases. In fact, in AIDS patients with PDH typical intracellular organisms can be seen directly on peripheral blood smears (buffy coat preparations) up to 50% of the time. Bronchoalveolar lavage also has a very high yield, both by direct smear and by culture in AIDS patients with high burden of organisms. Bronchoalveolar lavage offers the additional advantage of ready diagnosis of other opportunistic infections that are usually in the differential diagnosis, including *P. carinii* pneumonia.

Antigen Detection. Another approach to diagnosis of fungal infections is the use of the ultrasensitive assays for fungal antigens. This test is very useful when the burden of the infection is high. Detection of the histoplasma antigen in body fluids permits rapid diagnosis of PDH. The specificity of antigen detection is greater than 98%. There is, however, known cross reactivity with the other endemic fungi such as African histoplasmosis, blastomycosis, paracoccidioidomycosis, and *Penicillium marneffii* infection [32]. Sensitivity is lower in patients who are moderately immunosuppressed. However, the high density of organisms in AIDS patients with PDH makes antigen testing extremely useful in that setting.

Antigen testing is more sensitive on urine than serum. It is positive in either urine or serum in up to 95% of patients with PDH complicating AIDS [30]. Levels of histoplasma polysaccharide antigen in urine and serum also are useful for following the course of treatment and for predicting relapses [33], [34]. The test is done reliably in a single reference laboratory in Indianapolis, which makes it a "send out" for many institutions.

Treatment

Severe cases of PDH in non-AIDS patients are best treated promptly and aggressively with amphotericin B to a total dose of 40 mg/kg. Itraconazole (400 mg/day in a single daily dose) can be used successfully for patients with mild to moderate disease [35]. Sequential therapy for severely ill patients with amphotericin B to clinical improvement followed by 6 months of itraconazole is also being used but is not well studied.

AIDS patients are treated differently from other PDH cases. Relapse is expected if treatment is stopped. All patients require induction therapy to control symptoms and then maintenance therapy. Amphotericin B is used initially for all moderately and severely ill patients. After clinical response, treatment is changed to itraconazole (400 mg daily for 6 or more weeks, then 200 mg daily until sustained recovery of the immune system) [36], [37]. There is emerging data that support the discontinuation of maintenance Itraconazole in patients with HIV who recover their immune system [38].

Itraconazole can be used from onset for mild cases[39]. Other maintenance strategies for those intolerant of itraconazole include weekly amphotericin B infusions or fluconazole at doses of at least 400 mg/day. Ketoconazole therapy is ineffective for maintenance therapy, and fluconazole even at high daily doses (400 to 800 mg) is less effective than itraconazole [40]. In a recent double blind, randomized trial liposomal amphotericin B (L-AMB) was somewhat more effective than AMB in both time of response and survival, although the differences were not statistically significant [41]. In this study both types of AMB were used for fourteen days before switching to oral itraconazole therapy.

BLASTOMYCOSIS

Blastomycosis is an illness caused by the thermal dimorphic fungus *Blastomyces dermatitidis.* The spectrum of disease ranges from the asymptomatic acquisition of the fungus to a rapidly progressive and life-threatening respiratory or disseminated illness.

Epidemiology

Blastomycosis is most common in the central and south-central United States [42]. The proposed endemic area includes much of the central, south central, and southeastern United States, beginning near the Minnesota–North Dakota border and extending eastward and southward. The southeastern limit extends to South Carolina but not to Florida. This area overlaps most of the endemic area of histoplasmosis [1]. The northern limit, however, extends further. Northern Minnesota and northern Wisconsin and also the adjacent Canadian provinces of Ontario and Manitoba are endemic for blastomycosis but are free of histoplasmosis [43], [44] .

Similar to *H. capsulatum, B. dermatitidis* is a soil-dwelling fungus. Infection occurs by inhalation of airborne spores. Infected individuals develop a positive blastomycin skin test reaction or the in vitro correlate of delayed hypersensitivity. As with histoplasmosis, isolated microfoci of high infectivity exist in a large endemic area. Outbreaks often occur during activities such as hunting, camping, or canoeing in wooded or swampy environments [45]; these are most common when soil temperatures have been increasing for several days and when there is rain on the day of exposure. For sporadic cases, residence close to water in a highly endemic area and recent excavation activity are risk factors [46]. Dogs are also susceptible to blastomycosis. Canine blastomycosis is a well-recognized entity in veterinary practice in the endemic areas. The recognition of canine cases in a community should alert physicians that human blastomycosis may be present in their geographic area [47].

Pathogenesis

At ambient temperatures, the fungus grows as an aerial mycelium. When foci of actively growing blastomyces are disturbed, small 2- to 5-μm spores become airborne and an infective aerosol is formed. These infecting particles then may be inhaled by humans or by other mammals disturbing the site. Some spores may escape the nonspecific defenses of the lung and reach the alveoli.

The initial inflammatory response is neutrophilic. As the organism converts to the parasitic yeast form and begins to multiply, large numbers of yeast are seen, surrounded by neutrophils. Following this macrophages increase in number. Eventually, as specific cellular immunity develops, there are giant cells and well-formed epithelioid granulomas. In contrast to histoplasmosis, the neutrophilic component of the inflammatory response does not disappear completely, and the histopathologic examination often shows a mixed pyogenic and granulomatous response, even in chronic cases [2] .

It would be misleading, however, to think that there is a "characteristic" tissue response in blastomycosis. Occasionally, the neutrophilic component is minimal and the granulomas are noncaseating, producing a picture similar to that of sarcoidosis. In contrast, granulomas are sometimes entirely absent in overwhelming infections. The entire inflammatory reaction consists of neutrophils, and the histopathologic picture mimics that of bacterial infection.

The histopathologic response in cutaneous blastomycosis is striking. The stratified squamous epithelium becomes markedly hyperplastic, with exaggerated downgrowth of the rate pegs. Within these fingerlike projections are a number of microabscesses. The same hypertrophic tissue response is seen when the disease involves the oropharynx or the larynx. The histopathologic appearance may superficially resemble carcinoma. The characteristic organisms, seen best with special stains, provide a diagnosis.

The initial inflammatory and immune response may confine the infection to the lungs and the hilar lymph nodes. It is likely (but not proven) that self-limited early fungemia does not occur as often as it does in histoplasmosis. In some instances, however, the organism spreads beyond lung and the hilar nodes. Dissemination is usually to skin, bones, prostate, and meninges but can be seen in any organ [2], [48] .

The incubation time for blastomycosis is longer than for histoplasmosis and more variable. In the Eagle River outbreak, in which time of exposure was short and precisely defined, the median incubation period was 45 days, with a range of 21 to 106 days [49], [50].

Clinical Manifestations

The portal of entry is almost always the lung, and the primary illness is a lower respiratory infection. Some patients have an acute illness that resembles bacterial pneumonia, in contrast to acute pulmonary histoplasmosis, which more closely mimics influenza. The onset of symptoms is abrupt, with high fever and chills, followed by cough that rapidly becomes productive of large amounts of mucopurulent sputum. Pleuritic chest pain may occur.

This acute onset is common in an outbreak setting, but also may be seen in sporadic cases. In most sporadic cases, however, the onset of clinical symptoms is more gradual. The patient presents with a low-grade fever, productive cough, and weight loss [48] [51]. Lung cancer or tuberculosis are highest in the differential diagnosis, rather than bacterial pneumonia, depending on the roentgenographic findings.

Physical examination is usually unremarkable except for fever. Auscultation of the chest in patients who have segmental or lobar infiltrates may show crackles and focal consolidation; more often the physical examination is negative. Skin lesions are highly variable in appearance, ranging from subcutaneous nodules and abscesses to papules to ulcers with heaped up borders mimicking squamous cell cancers. Perhaps the most characteristic lesion has irregular borders and a crusted surface, varying in size from 1 to 10 or more centimeters. Skin lesions may be single or multiple and may occur in crops of several new lesions daily or every few days if the disease is rapidly disseminating.

Routine laboratory tests are seldom helpful. In cases resembling acute bacterial pneumonia, the white blood cell count is elevated, and frequently there is a shift to the left toward earlier forms in the granulocyte series.

There is no "characteristic" chest roentgenographic pattern in blastomycosis [52]. Lesions may vary from single or multiple round densities throughout both lung fields to segmental or lobar consolidation. Severe pulmonary infections can present with diffuse infiltrates, nodular, interstitial, or even alveolar (Figure 3). The diffuse alveolar infiltrates are identical to acute lung injury (as in acute respiratory distress syndrome) of diverse cause. Mass-like perihilar infiltrates, especially on the right side, are common and are often misinterpreted as neoplastic. On the lateral chest roentgenogram, the mass-like infiltrate is usually behind the hilum, in the apical-posterior segment of the lower lobe. Hilar lymph node involvement may occur but is not nearly as common as in histoplasmosis. Cavities may occur during the acute phase of the illness and usually close during successful treatment. Unlike histoplasmosis, calcification due to healed blastomycosis is rare.

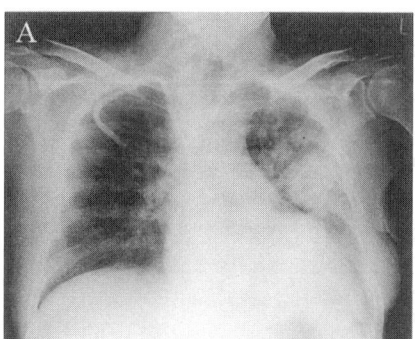 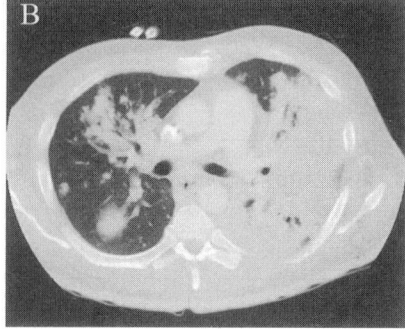

Figure 3
47 year old man who failed antibiotic therapy for a community acquired pneumonia, presenting with a rapidly progressive respiratory failure. Bronchoalveolar lavage recovered Blastomyces dermatitidis.

Extrapulmonary spread of the fungus may occur during the acute, symptomatic phase of the illness. In some instances, only the distant lesion (usually skin or bone) is symptomatic.

The skin and the bony skeleton are the most common sites of symptomatic extrapulmonary spread. The prostate gland, meninges, oral pharynx, larynx, and abdominal viscera, including the liver and the adrenal glands, are involved less frequently [2].

Blastomycosis can present as a progressive infection in patients with T-cell defects, including organ transplant recipients and other patients being treated with high-dose glucocorticoids and other immunosuppressive therapy for malignant and nonmalignant disorders. As with histoplasmosis, the disease can often be cured with amphotericin B in patients with intermediate degrees of immunosuppression.

Blastomycosis is much less common in AIDS and other T-cell-deficient conditions than are histoplasmosis and coccidioidomycosis. This is probably because exposure to this fungus, while immunosuppressed, is less common and because there is a smaller reservoir of patients with remote healed infection waiting to relapse should T-cell function markedly decline [53].

Blastomycosis can also occur in AIDS, usually with a CD4 count below 200/μl. Infection is particularly severe, and cure is not likely. Maintenance therapy is required for those who respond to initial treatment. Patients with AIDS are likely to progress with widespread dissemination of the infection with multi-organ involvement. Patients may present with sepsis like picture. Meningitis or brain abscess are common in this setting and it is associated with a high early mortality [54].

Diagnosis

The easiest and most rapid method of diagnosis is examination of expectorated sputum or aspirated pus after 10% potassium hydroxide digestion [2]. The characteristic large (8- to 20-μm) organism is easily identified. The yeast is single budding, with a broad neck of attachment between the parent and the daughter cells. The wall is thick and is double refractile, and there are multiple nuclei. Other direct fungal stains including periodic acid Schiff (PAS), calcoflour white, and silver stains are more sensitive. Another sensitive technique for rapid diagnosis of blastomycosis on direct sputum smears is cytologic analysis with the standard Papanicolaou stain (Figure 4). The direct techniques are probably complementary and examining multiple sputum samples increases diagnostic yield [55].

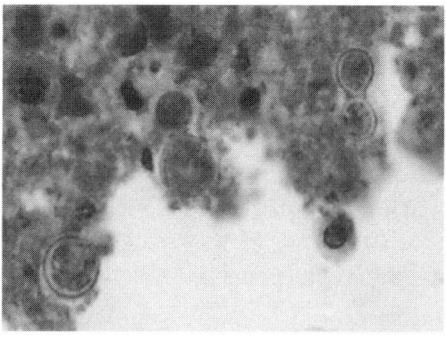

Figure 4
Papanicolaou stain of a sputum specimen from a
patient with rapidly progressive respiratory failure after
a community acquired pneumonia, showing the large
budding yeast with broad neck of blastomyses.

Bronchoscopy is useful when patients are unable to expectorate adequate sputum, and when urgent diagnosis is needed because of rapid pace of the illness. In one study, bronchoscopy (specimens obtained included bronchoalveolar lavage (BAL) in 64% and bronchial washings in all) was diagnostic in 92% of patients when culture results were included in the final analysis [56].

For patients who are acutely ill or have an ARDS-like picture, rapid diagnosis is crucial and can only be achieved by direct examination of respiratory secretions. If direct sputum smears are negative or not possible, then bronchoscopy should be done urgently with BAL and bronchial washings sent for both direct fungal stains (some combination of KOH, calcofluor white, silver stain, PAS and cytology preparation) and for culture [57]. The cytology laboratory should be informed whenever there is high clinical suspicion of infection. Cellblocks of concentrated BAL fluid can be done to maximize the yield of the submitted specimens.

Histopathologic examination of biopsy material is also an excellent way to establish the diagnosis. The decision whether or not to perform transbronchial biopsies at the time of initial bronchoscopy will likely depend on contraindications in any given patient that give added risk beyond risk of BAL. This is particularly true in critically ill patients. If blastomycosis is being considered in this setting, bronchoscopy with BAL can be the initial procedure, reserving transbronchial biopsy for cases with no diagnosis from a safer and easier first procedure. Standard hematoxylin and eosin stains do not stain the fungus and special stains are required. The periodic acid-Schiff stain preserves morphological detail, but silver stains are more commonly used and likely have better sensitivity.

Identification of the fungus by culture is not difficult, but it is slower. Growth may occur as early as 5 to 7 days but often takes several weeks. Exoantigen testing can provide positive identification as soon as good growth is established. Formerly, positive identification required conversion of the mycelial culture to the yeast phase of growth, adding 1 or more weeks of delay.

Serologic testing is of limited value in diagnosis of blastomycosis. They are positive in about a quarter of cases. Most cases of blastomycosis are diagnosed by smear, culture, and histopathology rather than by serological tests.

Treatment

Severely ill patients with pulmonary blastomycosis require immediate and aggressive treatment with amphotericin B.

Itraconazole is highly effective for blastomycosis and is the treatment of choice for most patients with pulmonary and nonmeningeal disseminated disease; oral therapy with 400 mg/day for 6 months successfully treats most patients with mild to moderate pulmonary disease, skin disease, and bone disease.

Amphotericin B (dosing as described for histoplasmosis; total cumulative dose 2000 mg) is preferred for a small minority of severely ill patients, including all patients with diffuse infiltrates and severe gas exchange abnormalities. Patients with edematous lobar pneumonia (bulging fissures), extremely toxic patients, and patients who are rapidly disseminating should all receive AMB. For these severe infections, sequential therapy with AMB to clinical improvement (usually 500 to 1000 mg total dose) followed by 6 months of oral itraconazole is often used and is effective though not well studied. This approach is also used for AIDS patients. Patients with AIDS and blastomycosis are not permanently cured. Life-long maintenance therapy is needed after induction therapy to improve symptoms.

Meningeal blastomycosis is always treated with systemic amphotericin B therapy in standard doses. L- AMB achieves higher brain tissue levels in animals and may be preferred. Fluconazole is overall less potent for blastomycocis than itraconazole but penetrates CNS much better. High dose fluconazole (often in combination with L-AMB) has been used for central nervous system blastomycosis. Voriconazole is theoretically attractive but has not been studied. Intracisternal amphotericin B has been used anecdotally in addition to systemic therapy for selected patients, but it is uncertain whether there is additional benefit [58], [59] . Intracisternal therapy is used less often now that there is a wider range of therapy.

COCCIDIOIDOMYCOSIS

Coccidioidomycosis is the illness caused by the tissue-dimorphic fungus *Coccidioides immitis.* Although most infections are mild and self-limited, the spectrum of illness includes life-threatening pulmonary disease and widely disseminated systemic disease with a high mortality rate. Differences between it and histoplasmosis and blastomycosis include different endemic areas, higher frequency of meningeal infections, and poorer response to all antifungal therapy, including amphotericin B.

Epidemiology

The endemic area for coccidioidomycosis in North America is the southwestern United States and the contiguous areas of northern Mexico. The endemic area of the United States includes central and southern California and extends eastward to Arizona, New Mexico, and western Texas.

In nature, the fungus grows as an aerial mycelium with septate hyphae. Alternating cells form thick-walled barrel-shaped structures called *arthroconidia,* with empty cells in between. When a natural site is disturbed, the mature arthrospores easily detach and become airborne, producing an infective aerosol.

The risk of infection is greatest during the hot dry summers. Strong winds can carry the arthrospores for long distances. A huge wind storm that blew north from the San Joaquin Valley in 1977 caused a major outbreak of coccidioidomycosis in Sacramento, far north of the usual endemic area [60]. Not surprisingly, occupations and activities with exposure to the soil carry the greatest risk for infection, including construction work, farm labor, and working on archeological digs [61].

Pathogenesis

After inhalation, some arthrospores evade the nonspecific lung defenses and reach the alveoli, where germination begins. The arthrospores develop into spherules, the tissue phase of the fungus. Spherules are large, round, thick-walled structures that vary in diameter between 10 and 80 μm. Reproduction of the fungus occurs within the spherule. The cytoplasm of the spherule undergoes progressive cleavage, forming numerous endospores within the spherule. Once a spherule matures, it bursts and releases the endospores into the surrounding tissues. Each endospore can become a new spherule and thus repeat the process.

The initial inflammatory response to inhaled arthrospores is neutrophilic. Resident alveolar macrophages also phagocytose the arthrospores and prime specific T lymphocytes, which multiply, recruit more macrophages, and arm the macrophages, engaging specific cell-mediated immunity. Even though many well-formed granulomas are seen, the neutrophilic inflammatory exudate does not disappear. Histopathologically, there is a mixed granulomatous and suppurative reaction more similar to blastomycosis than to histoplasmosis [3]. Granuloma formation is important for successful limitation of the infection. Good outcome correlates with preponderance of well-formed granulomas.

Most primary infections are asymptomatic or relatively mild. The fungus usually remains localized to the lung and hilar lymph nodes. Dissemination occurs in less than 1% of patients. Hematogenous spread can affect many tissues, including the skin, bones, lymph nodes, visceral organs, and meninges [62]. Meningitis is the most feared clinical syndrome with an ominous prognosis [63].

Clinical Manifestations

Except for rare instances of inoculation coccidioidomycosis, the portal of entry is the lung. About 60% of individuals with primary pulmonary infection remain totally asymptomatic. In the remaining 40%, the spectrum of disease ranges from a mild, influenza-like respiratory illness to a severe, life-threatening pneumonia [62].

The clinical symptoms and their severity are variable. Common symptoms include cough, fever, and pleuritic chest pain. Cough may be nonproductive, or there may be small amounts of mucopurulent sputum. True rigors are not common. Headache, common during the acute phase of the illness, is nonspecific. Severe headache is always worrisome, however, because coccidioidal meningitis often becomes clinically apparent during the early part of the illness. If meningitis is suspected, a lumbar puncture should be performed immediately.

Several dermatologic aspects of acute coccidioidomycosis are important. A mild nonspecific so-called toxic rash occurs in many patients [61], [64]. It is an erythematous macular rash that occurs early during the illness, before the skin test turns positive. Erythema nodosum and erythema multiforme are other skin manifestations of primary coccidioidal infection. Together with fever and arthralgias, these skin lesions are part of a variable symptom complex first recognized by locals in the San Joaquin valley of south central California and labeled "valley fever".

More than 75% of patients with primary coccidioidomycosis have an abnormal chest roentgenogram. The most common roentgenographic abnormality is a single or multiple areas of patchy pneumonitis. The ipsilateral hilar nodes are enlarged in about 25% of patients [65] . Hilar adenopathy may also be seen without recognizable parenchymal disease (Figure 5).

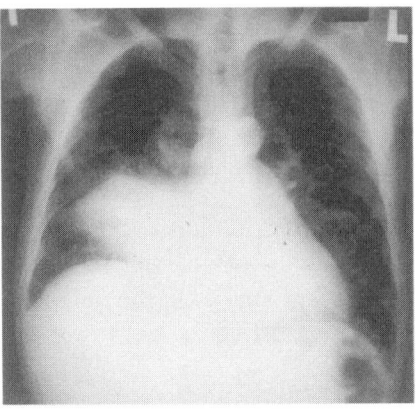

Figure 5
Acute pulmonary Coccidioidomycosis presenting
as an acute community acquired pneumonia.

This primary complex usually heals rapidly. Necrosis in the center of a pneumonic lesion may produce cavitation [66]. This is often accompanied by minor hemoptysis, which may be alarming but is seldom life-threatening. Hemoptysis as a late complication is uncommon but can be life-threatening, in contrast to the minimal bleeding that is often seen as the cavity forms. Another complication is rupture of the cavity with development of a pyopneumothorax.

In rare instances, primary pulmonary coccidioidomycosis is not self-limited but progresses within the lung. Symptoms are fever, cough, and weight loss. The chest roentgenogram shows progression of the infiltrate and variable involvement of the hilar nodes[65]. This form of pulmonary coccidioidomycosis is dangerous and augers impending dissemination. Most of the patients with progressive pulmonary disease are either immunosuppressed or belong to groups at high risk for dissemination.

The primary pulmonary infection usually either resolves completely or stabilizes. Rarely does a patient die when the disease is restricted to the lung. In some individuals, however, the fungus spreads widely throughout the body, resulting in a systemic infection known as disseminated coccidioidomycosis.

Patients receiving glucocorticoid or cytotoxic or newer immune modulating therapy for malignant or nonmalignant diseases are at risk of dissemination. This is especially true for recipients of renal and other organ transplant and patients with AIDS [67], [68]. The excess risk of coccidiodomycosis in organ transplant recipients in highly endemic areas has led to targeted prophylaxis to prevent re-activation whenever there is a history of coccidiodal infection or positive serologic results on pre-transplant screening [69]. There are other well-recognized risk factors for dissemination. Race and ethnicity are important. Disseminated coccidioidomycosis is more likely in blacks, Filipinos, and native Americans than in whites. Male gender is also a risk factor, as is diabetes mellitus. The very young and the very old are more likely to have dissemination[62]. There is much anecdotal information suggesting that coccidioidomycosis during the third trimester of pregnancy may be a severe illness with rapid dissemination.

Dissemination from the primary pulmonary focus tends to occur early, usually within a few months after a symptomatic pulmonary infection. In some patients, however, the findings of disseminated disease are the first manifestations of coccidioidomycosis, presumably because the preceding pulmonary infection was sub-clinical.

Dissemination may involve any organ in the body. The skin is one of the most common sites of dissemination and is involved in most patients some time in the course of the disease. Involvement of the bones is the next most common manifestation of disseminated coccidioidomycosis. Osteomyelitis may be either the sole evidence of extrapulmonary spread or part of a more widespread dissemination. Bone disease is usually restricted to one or two sites, but occasionally as many as eight separate lesions may be present.

Meningitis is the most dreaded complication of coccidioidal dissemination. Between one third and one half of all patients with disseminated disease have meningitis, frequently as the only obvious extrapulmonary site. The onset of meningitis may be subtle, with only mild headache and minimal alteration of mental functions. Striking boardlike nuchal rigidity, as in purulent meningitis, is seldom seen [63]. In fact, the findings of meningitis can be so minimal that all patients with dissemination at other sites should have a diagnostic lumbar puncture to exclude meningitis. Involvement of the base of the brain is characteristic. As the disease progresses, an exudate frequently obstructs the aqueduct of Sylvius and the foramina of the fourth ventricle, producing hydrocephalus. When obstruction occurs, the patient's clinical condition suddenly worsens, with diminished level of consciousness and the development of papilledema. The cerebrospinal fluid shows characteristics of chronic meningitis: predominantly mononuclear cell pleocytosis, increased protein, and decreased glucose. Occasionally, eosinophils are present in the cerebrospinal fluid. If present, they are a valuable clue to the possible coccidioidal nature of the chronic meningitis.

When coccidioidomycosis complicates HIV infection, the severity depends on the residual immune competence of the host. With near-normal CD4 lymphocyte counts, coccidioidomycosis is not significantly different from the disease seen in normal hosts. When the CD4 count falls below 250 cells/mL, disseminated disease tends to be severe and rapidly progressive. Patients usually have high fever, complain of dyspnea, and are hypoxemic; chest roentgenograms often show diffuse reticulonodular infiltrates with nodules 5 mm or greater in diameter (Figure 6). Diffuse macronodular pulmonary infiltrates are present in less than one per cent of non-AIDS patients with disseminated coccidioidomycosis, but in up to 50% of advanced AIDS patients with this condition. Meningeal disease is present in up to 25% of the patients [70], [71] .

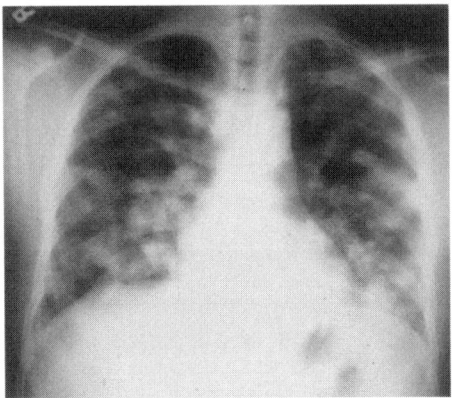

Figure 6
Disseminated Coccidioidomycosis in an HIV infected patient.

Diagnosis

Mycologic Studies. Direct examination of sputum and other respiratory specimens (or pus from a non-pulmonary site) may reveal the diagnostic spherules. Direct smears have highest utility in patients who produce copious sputum or have multi-lobar infiltrates [72] . Bronchoscopy is often performed in selected cases. In one study bronchoscopy was diagnostic in 69% of patients (compared to 32% for sputum stains and cultures) when patients with solitary pulmonary nodules on chest radiograph were excluded from analysis [73]. This study also showed usefulness of a post-bronchoscopy sputum and equivalent sensitivity for Papanicolaou and silver staining. The airway can be examined at the time of bronchoscopy and may be abnormal, providing clues to the diagnosis [74].

Bronchoscopy is typically performed in patients who are immunosuppressed and severely ill, especially if they have diffuse infiltrates on chest radiograph. Multiple infections often co- exist, adding additional value to diagnostic bronchoscopy early in the course of illness [75], [76] . Bronchial washings and bronchoalveolar lavage fluid should be sent for cytology, fungal stains, and culture. In a recent study of an AIDS patient in Phoenix, Arizona, the Papanicolaou stain was the most useful direct test (when compared to KOH and calcofluor white) for rapid diagnosis of pulmonary coccidioidomycosis and was even positive in two patients with negative cultures [77]. Histopathologic examination of biopsy material is extremely helpful. When mature spherules (visible on standard hematoxylin and eosin stained tissue sections) are seen, the diagnosis is secure. More commonly, only endospores, immature spherules or spherule fragments are present. Therefore fungal stains such as a silver stain should always be used in addition to hematoxylin and eosin staining. In one study, transbronchial biopsy yielded a specific tissue diagnosis of coccidioidomycosis in eight of eight patients.

Cultural identification of the fungus is not difficult but is hazardous to laboratory personnel. Isolation should be attempted only under rigid biohazard protection. Traditional laboratory methods for identifying culture isolates require conversion of mycelial-phase cultures to the tissue phase either by animal inoculation or directly by the use of slide cultures. Now immunodiffusion tests are performed directly on the supernatants of liquid mycelial-phase cultures. This method of identification (called exoantigen testing) is safer, simpler, and faster [78]. Positive identification of a coccidioidal isolate can sometimes be made by day 5, although it usually takes longer.

Serologic Studies. Because cultural identification is slow and even somewhat dangerous, serologic tests have been developed that facilitate rapid diagnosis [3] [79]. A tube precipitin tests for detection of Ig M antibodies is positive in 90% of patients by the third week (negative only in very mild infections). Because the test usually reverts to negative within 3 months, it is quite specific for recent infection[3]. Currently, an immunodiffusion test for IgM has largely replaced the tube precipitin test. The immunodiffusion test measures the same antibodies, but it is easier to perform.

The most important serodiagnostic test is the complement fixation (CF) test. CF antibodies are of the IgG class and appear later than IgM antibodies. In most symptomatic patients, the CF test is positive by 2 months and remains positive for several months or longer [79]. The test is highly specific but is not sensitive. Most asymptomatic skin test converters never have CF titers over 1:8, which is the threshold for a positive result. Most symptomatic patients have titers of 1:8 or 1:16. Titers of 1:32 or higher are generally associated with more severe infections and poorer prognosis. In the classic studies of Smith and colleagues [79], many patients with these high titers either had already undergone or were about to undergo dissemination. However, other patients with disseminated coccidioidomycosis did not have high titers. Also the cutoff of a 1:32 CF titer as a harbinger of dissemination never transferred perfectly to other laboratories that did not use the same method or the same antigen. A single CF titer, no matter how high, should never be used to make a diagnosis of disseminated coccidioidomycosis. Nonetheless, a steadily rising titer should raise the suspicion of disseminated coccidioidomycosis and prompt further tests (including bone scan, spinal tap, or both when appropriate) to better define the extent of disease.

Treatment

Because dissemination is more likely in immunosuppressed patients, in diabetics, and in certain racial and ethnic groups, it may be prudent to treat patients in high-risk groups during the primary infection, before dissemination takes place. In the past some authorities recommended a treatment course to a total dose of 500 to 2000 mg of amphotericin B [80]. Similarly, many experts believed that all patients with pulmonary disease that is severe or persists beyond a few weeks should receive amphotericin B to approximately the same total dose to prevent local pulmonary progression and to prevent dissemination. In current practice, many such patients (and also less symptomatic patients with pulmonary coccidioidomycosis of shorter duration) are often given fluconazole for 3-6 months, reserving AMB for patients with diffuse infiltrates and women in the third trimester of pregnancy. These recommendations are based on expert opinion and observational studies.

Amphotericin B is likely the best treatment for persistent pulmonary coccidioidomycosis. Because of their lesser toxicity, oral azoles are often tried. About two thirds of patients have clinical improvement with azole therapy, but many relapse when the course of treatment is finished. Ketoconazole was used first. Currently fluconazole and itraconazole are being used. Voriconazole will likely be evaluated in the future.

Disseminated coccidioidomycosis requires prompt and aggressive treatment. Unfortunately, amphotericin B is not as effective for disseminated coccidioidomycosis as it is for disseminated histoplasmosis or blastomycosis. The standard dose of amphotericin B is 2500 to 3000 mg given over many weeks or months. If necessary, much larger total doses may be given [81]. Daily doses of amphotericin B (usually 40 to 50 mg) are given while the patient is acutely ill. When the patient stabilizes, frequency should be reduced to three times weekly. Currently disseminated disease without CNS involvement should be treated with fluconazole or itraconazole first, especially in mild to moderate cases. AMB should be reserved for severe disease or treatment failure.

Fluconazole and itraconazole are now azoles of choice for nonmeningeal disseminated coccidioidomycosis. Neither is perfect for difficult cases for which even amphotericin B is often only suppressive. Long-term therapy is often required, extending to years or even indefinitely. Fluconazole has the advantage of better absorption, less gastrointestinal upset, and better penetration of the central nervous system. In a recently published randomized controlled trial, oral fluconazole and itraconazole were compared for treatment of non-meningeal coccidioidomycosis. Soft tissue dissemination responded best. Overall, itraconazole was somewhat more effective than fluconazole, producing response in 63% of the patients vs. 50% response in fluconazole treated patients (p = 0.08). Among patients with skeletal infections, itraconazole was clearly superior, (p=0.05) [82]. Some difficult cases of bone, lymph node, and soft-tissue coccidioidomycosis may be best managed with surgical drainage of focal abscesses, a 1000 to 2000 mg course of amphotericin B, and a prolonged course of itraconazole or fluconazole.

As might be expected, the treatment of disseminated coccidioidomycosis in AIDS is particularly difficult. Because of the rapid tempo of the disease, amphotericin B should be used initially, especially if the patient is severely ill. If the clinical course stabilizes, it is reasonable to switch to fluconazole for long-term suppression. Prognosis is poor. Even with prompt diagnosis and treatment, up to 40% of severely immunosuppressed patients die during the initial hospitalization. Other patients, usually with lesser degrees of immunosuppression, respond well to treatment[70], [71] .

Meningeal coccidioidomycosis is a major therapeutic challenge. The standard therapy in the past included a course of 2000 to 3000 mg systemic amphotericin therapy plus intensive and lengthy intrathecal (by lumbar or cisternal route) AMB therapy [63]. Intrathecal (or, less commonly, intraventricular via surgically placed reservoir [83]) AMB in doses between 0.25 and 1 mg was injected two to three times weekly until symptoms and cerebrospinal fluid pleocytosis resolve. Even after the patient had apparently recovered fully and cerebrospinal fluid pleocytosis had resolved, most authorities recommended continued injections of amphotericin to prevent relapse, first weekly and then at longer intervals. Relapses were common, but, with careful management, lengthy remissions could be obtained.

Because of the toxicity of this once standard approach to coccidioidomycotic meningitis, fluconazole has been evaluated as primary therapy for stable patients and as suppressive therapy after initial response to amphotericin B for more severely ill patients. Most patients respond favorably to fluconazole and maintain good clinical function. Dosage is 400 to 600 mg/daily or even higher. Therapy has to be continued long term, likely indefinitely [84]. Recently anecdotal reports have shown favorable response to voriconazole and this agent will undoubtedly be tried in various forms of coccidioidomycosis, including meningitis. A drug with potency and wide spectrum of itraconazole but with tissue penetration like fluconazole seems especially attractive for an treatment resistant illness with high incidence of meningeal spread. However clinical data is sparse.

Severely ill patients with both nonmeningeal and meningeal disease were previously treated with intravenous and intrathecal amphotericin B. Now they are sometimes treated with intravenous amphotericin B for faster, more effective initial therapy of the nonmeningeal disease and with fluconazole to control the central nervous system infection. Amphotericin B is continued to clinical improvement and fluconazole indefinitely.

Newer antifungal agents are being developed; their potential role in coccidioidomycosis is uncertain. As mentioned voriconazole has some promise because it has better CNS penetration than itraconazole – and yet may retain the potency advantage of itraconazole over fluconazole which has been demonstrated in non-meningeal disseminated disease.

REFERENCES

1. Goodwin, R.A., Jr. and R.M. Des Prez, State of the art: histoplasmosis. Am Rev Respir Dis, 1978. 117(5): p. 929-56.

2. Sarosi, G.A. and S.F. Davies, Blastomycosis. Am Rev Respir Dis, 1979. 120(4): p. 911-38.

3. Drutz, D.J. and A. Catanzaro, Coccidioidomycosis. Part I. Am Rev Respir Dis, 1978. 117(3): p. 559-85.

4. Davies, S.F., M. Khan, and G.A. Sarosi, Disseminated histoplasmosis in immunologically suppressed patients. Occurrence in a nonendemic area. Am J Med, 1978. 64(1): p. 94-100.

5. Goodwin, R.A., Jr., et al., Disseminated histoplasmosis: clinical and pathologic correlations. Medicine (Baltimore), 1980. 59(1): p. 1-33.

6. Manfredi, R., et al., Histoplasmosis capsulati and duboisii in Europe: the impact of the HIV pandemic, travel and immigration. Eur J Epidemiol, 1994. 10(6): p. 675-81.

7. Edwards, L.B., et al., An atlas of sensitivity to tuberculin, PPD-B, and histoplasmin in the United States. Am Rev Respir Dis. Vol. 99. 1969. Suppl:1-132.

8. Hammerman, K.J., K.E. Powell, and F.E. Tosh, The incidence of hospitalized cases of systemic mycotic infections. Sabouraudia, 1974. 12(1): p. 33-45.

9. Wheat, L.J., et al., A large urban outbreak of histoplasmosis: clinical features. Ann Intern Med, 1981. 94(3): p. 331-7.

10. Wheat, L.J., et al., Disseminated histoplasmosis in the acquired immune deficiency syndrome: clinical findings, diagnosis and treatment, and review of the literature. Medicine (Baltimore), 1990. 69(6): p. 361-74.

11. Procknow, J.J., M.I. Page, and C.G. Loosli, Early pathogenesis of experimental histoplasmosis. Arch Pathol, 1960. 69: p. 413-26.

12. Zhou, P., G. Miller, and R.A. Seder, Factors involved in regulating primary and secondary immunity to infection with Histoplasma capsulatum: TNF-alpha plays a critical role in maintaining secondary immunity in the absence of IFN-gamma. J Immunol, 1998. 160(3): p. 1359-68.

13. Allendoerfer, R., G.P. Biovin, and G.S. Deepe, Jr., Modulation of immune responses in murine pulmonary histoplasmosis. J Infect Dis, 1997. 175(4): p. 905-14.

14. Allendorfer, R., G.D. Brunner, and G.S. Deepe, Jr., Complex requirements for nascent and memory immunity in pulmonary histoplasmosis. J Immunol, 1999. 162(12): p. 7389-96.

15. Allendoerfer, R. and G.S. Deepe, Jr., Intrapulmonary response to Histoplasma capsulatum in gamma interferon knockout mice. Infect Immun, 1997. 65(7): p. 2564-9.

16. Wood, K.L., et al., Histoplasmosis after treatment with anti-tumor necrosis factor-alpha therapy. Am J Respir Crit Care Med, 2003. 167(9): p. 1279-82.

17. Straub, M. and J. Schwarz, The healed primary complex in histoplasmosis. Am J Clin Pathol, 1955. 25(7): p. 727-41.

18. Wheat, J., Histoplasmosis. Experience during outbreaks in Indianapolis and review of the literature. Medicine (Baltimore), 1997. 76(5): p. 339-54.

19. Conces, D.J., Jr., et al., Disseminated histoplasmosis in AIDS: findings on chest radiographs. AJR Am J Roentgenol, 1993. 160(1): p. 15-9.

20. Davies, S.F., R.W. McKenna, and G.A. Sarosi, Trephine biopsy of the bone marrow in disseminated histoplasmosis. Am J Med, 1979. 67(4): p. 617-22.

21. Wheat, L.J., T.G. Slama, and M.L. Zeckel, Histoplasmosis in the acquired immune deficiency syndrome. Am J Med, 1985. 78(2): p. 203-10.

22. Johnson, P.C., R.J. Hamill, and G.A. Sarosi, Clinical review: progressive disseminated histoplasmosis in the AIDS patient. Semin Respir Infect, 1989. 4(2): p. 139-46.

23. Wheat, L.J., et al., Risk factors for disseminated or fatal histoplasmosis. Analysis of a large urban outbreak. Ann Intern Med, 1982. 96(2): p. 159-63.

24. Sarosi, G.A., et al., Disseminated histoplasmosis: results of long-term follow-up. A center for disease control cooperative mycoses study. Ann Intern Med, 1971. 75(4): p. 511-6.

25. Wheat, L.J., B.E. Batteiger, and B. Sathapatayavongs, Histoplasma capsulatum infections of the central nervous system. A clinical review. Medicine (Baltimore), 1990. 69(4): p. 244-60.

26. Hawkins, S.S., D.W. Gregory, and R.H. Alford, Progressive disseminated histoplasmosis; favorable response to ketoconazole. Ann Intern Med, 1981. 95(4): p. 446-9.

27. Vail, G.M., et al., Incidence of histoplasmosis following allogeneic bone marrow transplant or solid organ transplant in a hyperendemic area. Transpl Infect Dis, 2002. 4(3): p. 148-51.

28. Davies, S.F., Serodiagnosis of histoplasmosis. Semin Respir Infect, 1986. 1(1): p. 9-15.

29. Prechter, G.C. and U.B. Prakash, Bronchoscopy in the diagnosis of pulmonary histoplasmosis. Chest, 1989. 95(5): p. 1033-6.

30. Wheat, L.J., et al., Diagnosis of histoplasmosis in patients with the acquired immunodeficiency syndrome by detection of Histoplasma capsulatum polysaccharide antigen in bronchoalveolar lavage fluid. Am Rev Respir Dis, 1992. 145(6): p. 1421-4.

31. Salzman, S.H., R.L. Smith, and C.P. Aranda, Histoplasmosis in patients at risk for the acquired immunodeficiency syndrome in a nonendemic setting. Chest, 1988. 93(5): p. 916-21.

32. Wheat, J., et al., Cross-reactivity in Histoplasma capsulatum variety capsulatum antigen assays of urine samples from patients with endemic mycoses. Clin Infect Dis, 1997. 24(6): p. 1169-71.

33. Wheat, L.J., et al., Histoplasmosis relapse in patients with AIDS: detection using Histoplasma capsulatum variety capsulatum antigen levels. Ann Intern Med, 1991. 115(12): p. 936-41.

34. Wheat, L.J., et al., Effect of successful treatment with amphotericin B on Histoplasma capsulatum variety capsulatum polysaccharide antigen levels in patients with AIDS and histoplasmosis. Am J Med, 1992. 92(2): p. 153-60.

35. Dismukes, W.E., et al., Itraconazole therapy for blastomycosis and histoplasmosis. NIAID Mycoses Study Group. Am J Med, 1992. 93(5): p. 489-97.

36. Wheat, J., et al., Prevention of relapse of histoplasmosis with itraconazole in patients with the acquired immunodeficiency syndrome. The National Institute of Allergy and Infectious Diseases Clinical Trials and Mycoses Study Group Collaborators. Ann Intern Med, 1993. 118(8): p. 610-6.

37. Hecht, F.M., et al., Itraconazole maintenance treatment for histoplasmosis in AIDS: a prospective, multicenter trial. J Acquir Immune Defic Syndr Hum Retrovirol, 1997. 16(2): p. 100-7.

38. Goldman, M., et al., Safety of Discontinuation of Maintenance Therapy for Disseminated Histoplasmosis after Immunologic Response to Antiretroviral Therapy: AIDS Clinical Trials Group Study A5038. Clinical Iinfectious Diseases, 2004:38 (15 May). In press .

39.   Wheat, J., et al., Itraconazole treatment of disseminated histoplasmosis in
       patients with the acquired immunodeficiency syndrome. AIDS Clinical Trial
       Group. Am J Med, 1995. 98(4): p. 336-42.

40.   Wheat, J., et al., Treatment of histoplasmosis with fluconazole in patients with
       acquired immunodeficiency syndrome. National Institute of Allergy and
       Infectious Diseases Acquired Immunodeficiency Syndrome Clinical Trials
       Group and Mycoses Study Group. Am J Med, 1997. 103(3): p. 223-32.

41.   Johnson, P.C., et al., Safety and efficacy of liposomal amphotericin B compared
       with conventional amphotericin B for induction therapy of histoplasmosis in
       patients with AIDS. Ann Intern Med, 2002. 137(2): p. 105-9.

42.   Furcolow, M.L., et al., Prevalence and incidence studies of human and canine
       blastomycosis. 1. Cases in the United States, 1885-1968. Am Rev Respir Dis,
       1970. 102(1): p. 60-7.

43.   Kepron, M.W., et al., North American blastomycosis in Central Canada. A
       review of 36 cases. Can Med Assoc J, 1972. 106(3): p. 243-6.

44.   Tosh, F.E., et al., A common source epidemic of North American
       blastomycosis. Am Rev Respir Dis, 1974. 109(5): p. 525-9.

45.   Greenberg, S.B., Serious waterborne and wilderness infections. Crit Care Clin,
       1999. 15(2): p. 387-414.

46.   Baumgardner, D.J. and K. Brockman, Epidemiology of human blastomycosis in
       Vilas County, Wisconsin. II: 1991-1996. Wmj, 1998. 97(5): p. 44-7.

47.   Sarosi, G.A., et al., Canine blastomycosis as a harbinger of human disease. Ann
       Intern Med, 1979. 91(5): p. 733-5.

48.   Witorsch, P. and J.P. Utz, North American blastomycosis: a study of 40
       patients. Medicine (Baltimore), 1968. 47(3): p. 169-200.

49.   Klein, B.S., et al., Isolation of Blastomyces dermatitidis in soil associated with
       a large outbreak of blastomycosis in Wisconsin. N Engl J Med, 1986. 314(9): p.
       529-34.

50.   Klein, B.S., J.M. Vergeront, and J.P. Davis, Epidemiologic aspects of
       blastomycosis, the enigmatic systemic mycosis. Semin Respir Infect, 1986.
       1(1): p. 29-39.

51.   Abernathy, R.S., Clinical manifestations of pulmonary blastomycosis. Ann
       Intern Med, 1959. 51: p. 707-27.

52.   Laskey, W. and G.A. Sarosi, The radiological appearance of pulmonary
       blastomycosis. Radiology, 1978. 126(2): p. 351-7.

53.   Davies, S. and G. Sarosi, Clinical manifestations and management of
       blastomycosis in the compromised patient. In Warnock DW, Richard MD (eds):
       Fungal Infection in the Compromised Patient. New York: John Wiley & Sons,
       1982, 1982: p. 215-229.

54.   Pappas, P.G., et al., Blastomycosis in patients with the acquired immunodeficiency syndrome. Ann Intern Med, 1992. 116(10): p. 847-53.

55.   Trumbull, M.L. and T.M. Chesney, The cytological diagnosis of pulmonary blastomycosis. Jama, 1981. 245(8): p. 836-8.

56.   Martynowicz, M.A. and U.B. Prakash, Pulmonary blastomycosis: an appraisal of diagnostic techniques. Chest, 2002. 121(3): p. 768-73.

57.   Lemos, L.B., M. Guo, and M. Baliga, Blastomycosis: organ involvement and etiologic diagnosis. A review of 123 patients from Mississippi. Ann Diagn Pathol, 2000. 4(6): p. 391-406.

58.   Gonyea, E.F., The spectrum of primary blastomycotic meningitis: a review of central nervous system blastomycosis. Ann Neurol, 1978. 3(1): p. 26-39.

59.   Kravitz, G.R., et al., Chronic blastomycotic meningitis. Am J Med, 1981. 71(3): p. 501-5.

60.   Flynn, N.M., et al., An unusual outbreak of windborne coccidioidomycosis. N Engl J Med, 1979. 301(7): p. 358-61.

61.   Werner, S.B., et al., An epidemic of coccidioidomycosis among archeology students in northern California. N Engl J Med, 1972. 286(10): p. 507-12.

62.   Drutz, D.J. and A. Catanzaro, Coccidioidomycosis. Part II. Am Rev Respir Dis, 1978. 117(4): p. 727-71.

63.   Bouza, E., et al., Coccidioidal meningitis. An analysis of thirty-one cases and review of the literature. Medicine (Baltimore), 1981. 60(3): p. 139-72.

64.   Bayer, A.S., et al., Unusual syndromes of coccidioidomycosis: diagnostic and therapeutic considerations; a report of 10 cases and review of the English literature. Medicine (Baltimore), 1976. 55(2): p. 131-52.

65.   Bayer, A.S., Fungal pneumonias; pulmonary coccidioidal syndromes (Part I). Primary and progressive primary coccidioidal pneumonias -- diagnostic, therapeutic, and prognostic considerations. Chest, 1981. 79(5): p. 575-83.

66.   Winn, W.A., A long term study of 300 patients with cavitary-abscess lesions of the lung of coccidioidal origin. An analytical study with special reference to treatment. Dis Chest, 1968. 54: p. Suppl 1:268+.

67.   Rutala, P.J. and J.W. Smith, Coccidioidomycosis in potentially compromised hosts: the effect of immunosuppressive therapy in dissemination. Am J Med Sci, 1978. 275(3): p. 283-95.

68.   Cohen, I.M., et al., Coccidioidomycosis in renal replacement therapy. Arch Intern Med, 1982. 142(3): p. 489-94.

69.   Blair, J.E., D.D. Douglas, and D.C. Mulligan, Early results of targeted prophylaxis for coccidioidomycosis in patients undergoing orthotopic liver transplantation within an endemic area. Transpl Infect Dis, 2003. 5(1): p. 3-8.

70. Bronnimann, D.A., et al., Coccidioidomycosis in the acquired immunodeficiency syndrome. Ann Intern Med, 1987. 106(3): p. 372-9.

71. Fish, D.G., et al., Coccidioidomycosis during human immunodeficiency virus infection. A review of 77 patients. Medicine (Baltimore), 1990. 69(6): p. 384-91.

72. Warlick, M.A., S.F. Quan, and R.E. Sobonya, Rapid diagnosis of pulmonary coccidioidomycosis. Cytologic v potassium hydroxide preparations. Arch Intern Med, 1983. 143(4): p. 723-5.

73. Wallace, J.M., et al., Flexible fiberoptic bronchoscopy for diagnosing pulmonary coccidioidomycosis. Am Rev Respir Dis, 1981. 123(3): p. 286-90.

74. Polesky, A., et al., Airway coccidioidomycosis--report of cases and review. Clin Infect Dis, 1999. 28(6): p. 1273-80.

75. Mahaffey, K.W., et al., Unrecognized coccidioidomycosis complicating Pneumocystis carinii pneumonia in patients infected with the human immunodeficiency virus and treated with corticosteroids. A report of two cases. Arch Intern Med, 1993. 153(12): p. 1496-8.

76. Sobonya, R.E., et al., Detection of fungi and other pathogens in immunocompromised patients by bronchoalveolar lavage in an area endemic for coccidioidomycosis. Chest, 1990. 97(6): p. 1349-55.

77. Sarosi, G.A., et al., Rapid diagnostic evaluation of bronchial washings in patients with suspected coccidioidomycosis. Semin Respir Infect, 2001. 16(4): p. 238-41.

78. Standard, P.G. and L. Kaufman, Immunological procedure for the rapid and specific identification of Coccidioides immitis cultures. J Clin Microbiol, 1977. 5(2): p. 149-53.

79. Smith, C.E., M.T. Saito, and S.A. Simons, Pattern of 39,500 serologic tests in coccidioidomycosis. J Am Med Assoc, 1956. 160(7): p. 546-52.

80. Galgiani, J.N., et al., Practice guideline for the treatment of coccidioidomycosis. Infectious Diseases Society of America. Clin Infect Dis, 2000. 30(4): p. 658-61.

81. Bennett, J.E., Chemotherapy of systemic mycoses (first of two parts). N Engl J Med, 1974. 290(1): p. 30-2.

82. Galgiani, J.N., et al., Comparison of oral fluconazole and itraconazole for progressive, nonmeningeal coccidioidomycosis. A randomized, double-blind trial. Mycoses Study Group. Ann Intern Med, 2000. 133(9): p. 676-86.

83. Diamond, R.D. and J.E. Bennett, A subcutaneous reservoir for intrathecal therapy of fungal meningitis. N Engl J Med, 1973. 288(4): p. 186-8.

84. Dewsnup, D.H., et al., Is it ever safe to stop azole therapy for Coccidioides immitis meningitis? Ann Intern Med, 1996. 124(3): p. 305-10.

INDEX